Resource Conservation Technologies for Food and Environmental Security

NIPA® GENX ELECTRONIC RESOURCES & SOLUTIONS P. LTD.
New Delhi-110 034

About Authors

Dr. Sunil Kumar is the Scientist working in the Department of Agronomy, Bihar Agricultural University, Sabour. He did his B.Sc (Ag) from BHU, Varanasi and M.Sc (Ag) in Agrometrology from Govind Ballabh Pant University of Agriculture and Technology, Pantnagar as M.Sc. (Ag) in Agrometeorology. He accomplished his Ph.D. in Agrometeorology and Environmental Sciences from SHUATS, Prayagraj, U.P. He availed ICAR-Junior Research Fellowship in Physical Sciences for pursuing his M.Sc. (Ag) and CSIR-Junior Research Fellowship in Earth Atmosphere, Ocean and Planetary Sciences for Ph.D. He was honored with several awards like Young Scientist Award for outstanding contribution. He was awarded with Excellence in Teaching Award for outstanding contribution in Teaching. He was also awarded with best presentation award in the National Conference CAGSAS-2022 at BAU Sabour. He has been actively involved in teaching, research and extension at BAU Sabour since 2012. He has published several research papers in national and international journals of repute. He had also a significant contribution in the field of Aerosol and its impact on Crop productivity, Crop Simulation, Agromet Advisory Services, Remote Sensing and GIS at Bihar Agricultural University, Sabour.

Dr. D.R. Singh, born in the year 1964 completed his graduation from Gorakhpur University in 1986 and earned his post graduate and doctorate degree in the field of horticulture with specialization in Floriculture from Agra University in the year 1989 and 1992, respectively. He started his career as an Assistant Professor at ANGRAU, Hyderabad in the year 1994 and later joined ICAR as Senior Scientists in the year 2001. Dr. Singh has been known for his exemplary research work for Noni, an under-utilized horticultural crop during his vast tenure as Principal Scientist, Head and Director at ICAR-CIARI, Port Blair and Director, NRC for Orchids, Sikkim. Apart from developing various situation-based technologies, concepts, products and methodologies, he has given 22 varieties of horticulture crops, which has created a big impact on area and production and livelihood of farming community. During his service period of about 32 years, he has experience of working in various capacities such as Head of the Department, Director and Vice Chancellor. Dr. Singh has been chairman, vice president, and member of various national and international executive bodies, professional societies and task forces. During his tenure as Vice Chancellor of CSAUAT, Kanpur he brought laurels to university by getting NAAC B+ accreditation and 30th rank in NIRF 2023 among the agriculture and allied sector groups.Presently, as the Vice Chancellor of Bihar Agricultural University, Sabour; Dr Singh is committed to the development of the farming community of the state in days and years to come and shall focus on the improvement in the standards of agricultural education, research, extension and training so as to make it a state-of-art university.

Dr. Anil Kumar Singh is an accomplished agricultural scientist who has made significant contributions to the field of agricultural research. He started his career as a Scientist at the National Bureau of Plant Genetic Resources, Regional Station Srinagar (Jammu & Kashmir). Dr. Singh has held several key positions during his career, including Principal Scientist at the ICAR Research Complex for Eastern Region, Patna. He has served as the Head (Acting) of the Research Programme at Pusa, Bihar and as the Officer In-charge at NBPGR RS Srinagar (Jammu & Kashmir). Dr. Singh has an impressive publication record, with over 150 research articles, 10 books, and 196 other publications. Dr. Singh has taken on editorial responsibilities as the Editor-in-Chief of the 'Journal of AgriSearch' and e-kheti (an online publication). He has also served as the प्रधान संपादक of 'कृषि पंजूषा,' demonstrating his commitment to disseminating agricultural knowledge. Dr. Singh has received several prestigious awards and recognitions, including the Ganesh Shankar Vidyarthi Hindi Patrika Purskaar for 'Akshay Kheti' from ICAR, New Delhi. He is a recipient of the Fellow of the Indian Society of Agronomy, Fellow of the Indian Society of Plant Genetic Resources, and an Associate of the National Academy of Agricultural Sciences, New Delhi. Dr. Singh has also been honored with the Rajbhasha Gaurav Award by the Department of Official Languages, Ministry of Home Affairs, Government of India, for his outstanding technical book in Hindi. Dr. Singh has demonstrated his organizational skills by serving as the Chief Organizing Secretary for various conferences, including the International Conference on Rural in 2018 and National Conference in 2016.

Mr. Sushanta Sarkar has been working as Technical Officer under GKMS Project (Agrometeorology), Bihar Agricultural College, Sabour, Bhagalpur since November 2019. He received his M.Sc. degree in Agricultural Meteorology from the Bidhan Chandra Krishi Vishwavidyalaya, Mohanpur (WB) and ICAR-NET in 2017. He published more than 15 publications (research papers, journal articles, book chapters) and more than 10 Hindi bulletins.

Resource Conservation Technologies for Food and Environmental Security

Sunil Kumar
Scientist
Department of Agronomy
Bihar Agricultural University
Sabour, Bihar, India

D.R. Singh
Vice-Chancellor
Bihar Agricultural University
Sabour, Bihar, India

Anil Kumar Singh
Director Research
Bihar Agricultural University
Sabour, Bihar, India

Sushanta Sarkar
Technical Officer
GKMS Project (Agrometeorology)
Bihar Agricultural College
Sabour, Bihar, India

NIPA® GENX ELECTRONIC RESOURCES & SOLUTIONS P. LTD.
New Delhi-110 034

NIPA® GENX ELECTRONIC RESOURCES & SOLUTIONS P. LTD.

101,103, Vikas Surya Plaza, CU Block
L.S.C.Market, Pitam Pura, New Delhi-110 034
Ph : +91 11 27341616, 27341717, 27341718
E-mail:newindiapublishingagency@gmail.com
www: www.nipabooks.com

For customer assistance, please contact
Phone: + 91-11-27 34 17 17
Fax: + 91-11- 27 34 16 16
E-Mail: feedbacks@nipabooks.com

ISBN: 978-93-58876-27-7

Composed and Designed by NIPA®.

Preface

Natural resources are vital for meeting the food, livelihood and environmental security. It is now widely accepted that future of food, livelihood and environmental security depends upon the attention paid to conservation, sustainable development and management of natural resources. Research carried out over the last five decades have clearly demonstrated that resource conserving technological interventions, water harvesting and improved cultivation practices have the potential to improve utilization of natural resources and significantly enhance crop productivity in the country.

The chapters in this book are multidisciplinary and covering a wide range of information for better understanding of the diverged problems in the field of Natural Resource Management which includes Degradation of Resources, Climate Change, Bio-diversity, Remote Sensing and GIS, Soil and Water Conservation, Water Resource Development, Land Use and Land Cover, Watershed Management, Incorporation of pulse crops, Weed Management, Community Participation in Resource Management, Sustainable Agriculture and Food Security, Socio-Economical Aspects, Regional Experiences etc. Each chapter provides up-dated information, detailed content in the form of figures or tables and future challenges to achieve desired goals of food and environmental security through resource conservation.

We hope the book will be of immense importance for scientists, teachers, scholars, policy makers and extension agencies working in the area of natural resource conservation, and also for those who are interested in judicious management of natural resources for sustainable development to meet demands of the growing population.

Authors

Contents

List of Figures

List of Tables

1

Resource Conservation Technologies (RCTs): Needs and Future Prospects

1.1 Introduction

Broadcasting seed and harvesting meagre yields was a common method of ancient agriculture. High use of developed crop varieties, irrigation, chemical fertilizers, pesticides and farm mechanization is the main thrust of modern agriculture. Therefore, modern agriculture is an energy intensive farming system (Khan *et al.*, 2010). Though modern technology based agriculture is achieving success to meet food demand of increasing global population, yet it creates many environmental issues such as loss of soil fertility, salinization and water scarcity. Conventional agriculture is characterized by tillage operations which includes soil loosening and levelling, chemical fertilizer application, weed control and crop residue management. However, continuous use of conventional farm practices with conventional tillage and burning of crop residues has degraded the soil resources and intensified soil degradation by 67% with concomitant decrease in crop production capacity (World Resources Institute, 2000). Without a sustainable boost to crop productivity, agricultural sector will not be able to meet the rising demand of foods caused by increasing global population (Foresight, 2011). Today's real challenges for agriculture scientists are to increase crop productivity, decrease cost of cultivation and socioeconomic changes without hampering environment. So, there is a need to focus on resource conservation technologies (RCTs) which give emphasis on efficient use of natural resources which include soil, water, air, inputs and people. Now conservation agriculture based RCTs are being practiced by the farmers over 100 M ha area worldwide under different soil types, climatic conditions and geographic zones. RCTs are energy and input efficient and reduce emerging environment and soil health problems. The CA based RCTs include no- tillage or minimum tillage, crop residue management, direct seeding or bed planting and crop rotaion or intercropping which have the potential to improve crop productivity and soil fertility by increasing soil organic matter content (Bhattacharyya *et al.*, 2013a,b). The RCTs reduce use of fossil fuels and electricity resulting reduction in cultivation cost and reduces greenhouse gas (GHG) emissions resulting reduction in air pollution. RCTs also decrease soil erosion, degradation of the natural resource and labour requirement and increase crop yield and farm income.

1.2 Principles of CA

CA has four principles viz.:

(i) Seeding directly into untilled soil with minimum mechanical disturbance of soil to improve soil health.

(ii) Increasing SOM content using crop residues and cover crops. This method also protects soil surface form erosion, conserves water and soil nutrients and increase soil biological activity.

(iii) Diversification of crops by crop rotation or using intercrops to enhance system resilience and break pest and disease cycles.

(iv) CA avoids burning of crop residues, enhances soil organic carbon content, increases input use efficiency and also has the potential to reduce emissions of GHGs.

Different field experiments throughout the globe have shown that zero-tilled soil has more SOC content than tilled soil. Raised bed planting of crop generally reduces irrigation water and labour requirement without affecting crop productivity. Residue retention on the soil surface generally increases organic matter content in the soil and improves crop productivity significantly. Different conservation tillage practices along with efficient management of irrigation increases SOC.

1.3 Need of CA

- Now, one of the most important challenges facing by agricultural scientists is to increase food production to maintain food security and meet needs of increasing global population. However, increase in crop production can be achieved sustainably by minimizing negative impacts of the environment on crops and increasing income to improve the livelihoods of the farmers.
- The demand for food is increasing day by day not only to meet food security for increasing population, but also to provide nutritional security to the people.
- To maintain ecological balance and to make the resources available for present and future generation.

- As domestic and industrial needs are competing for surface and groundwater resources, water scarcity will be more severe.
- Agricultural land is shrinking due to urbanization. So proper maintenance of lands that are already in use for agriculture is mandatory to maintain food security.
- Costs of fossil fuels are increasing day by day, as a result cost of production also increasing directly or indirectly. Burning of fossil fuels increase emissions of GHGs resulting severe climatic events such as cyclones, flood, drought etc.
- The best way to achieve sustainable food production is to use of natural resources more efficiently.

Table 1.1: Some distinguishing features of conventional and conservation agriculture systems

Conventional agriculture	**Conservation agriculture**
Cultivating land, using science and technology to dominate nature.	Least interference with natural processes.
Excessive mechanical tillage and soil erosion.	No-till or drastically reduced tillage (biological tillage).
High wind and soil erosion.	Low wind and soil erosion.
Burning or removal of crop residues.	Surface retention of crop residues.
Water infiltration rate is low.	Water infiltration rate is high.
Use of ex-situ FYM/composts.	Use of in-situ organics/composts.
Green manuring (incorporated).	Brown manuring/cover crops (surface retention).
Kills weeds but also stimulates more weed seeds to germinate.	Weeds problem is more in the early stages but decrease with time.
Increased soil compaction.	N compaction in crop field.
Mono culture and less efficient rotations.	Diversified culture and more efficient rotations.
Heavy requirement of manual labour.	Due to mechanized operations manual labour requirement is less.
Poor adaptation to stress and yield loss is more under stress condition.	More resilience to stress and yield loss is less under stress condition.
Productivity gains are in declining order	Productivity gains are in incremental order

1.4 RCTs for CA

A. Zero tillage

Zero tillage (ZT) is the process where soil is not disturbed or turned for sowing of seeds or planting the crop plants. In the traditional field preparation, 3-4 tillage operations are done for ploughing which is both the time and cost consuming. So, adoption of zero tillage reduces cost of production and timely planting of crop is also ensured. Malik *et al.* (2002) mentioned that in case of early sowing of wheat under ZT, infestation of *Phalaris minor* which is an herbicide-resistant weed in wheat is comparatively less than when wheat is sown with conventional tillage practices. Effect of zero tillage on crop yields is highly variable. Zero tillage in combination with mulching decreases crop yields initially and but increases the yields over the subsequent time (Baudron *et al.*, 2011). Wider adoptions of zero tillage in wheat cultivation by the global farmers are being seen due to increase in crop yield and reduce in production cost.

Figure 1.1: Crop germination in zero tillage practice

B. Crop residue cover

Crop residues are the parts crop plants that are left in the field after harvesting. Each year, India alone produces 350-990 million tonnes of agricultural waste and over 130 million tonnes of paddy straw out of which half is used as fodder and the other half is discarded (Singh and Sidhu, 2014). Crop residues are the important sources of soil organic matter and maintain physical, chemical and biological properties of soil. Use of crop residue in rice-wheat cropping system increases rice and wheat yields by 13 and 8% respectively, decreases cost of cultivation by 5 and 3% and energy efficiency by 13 and 6%, respectively (PDFSR, 2011). However, decomposition rate and release of nitrogen from crop residues depends on soil types, climatic conditions and the C: N ratio of residues. Crop residues also act as mulching and protect the soil surface from erosion, reduce evaporation loss, maintain soil temperature and increase soil microbial activities.

Figure 1.2: Crop residue cover

C. Cover crops

Covering the soil surface with crop residues or cover crops is another fundamental principle of CA. After harvesting the crop, crop residues are left on the soil surface. But cover crops are grown if the gap is long between cultivation of two crops. In CA system, commercial crops like rice, wheat or maize are generally relayed by or intercropped with a cover crop like *Crotalaria* sp., *Brachiaria* sp., *Stylosanthes* sp. or *Cajanus* sp. Cover crop produces extra biomass that can be used as green manure or fodder for livestock. Huge biomass of cover crop acts as mulching on the soil. They not only improve the stability of the CA system by improving the soil health but also improve biodiversity in the agro-ecosystem. When cover crops are intercropped with commercial crops, they significantly increase radiation interception by the plant canopy. Huge vegetative cover also protects the soil surface from erosion by raindrops and reduces evaporation loss as they act as mulching. If legume crops are cultivated as cover crops they increase the soil fertility through nitrogen fixation. Decomposition of litter falls also adds organic matter to the soil.

Figure 1.3: Cover crop

D. Precision farming

Precision agriculture, also known as precision farming, is a type of farming that takes into account in-field variability and implements planting, fertiliser management, and plant protection measures based on the local conditions of a given field. As a result, agricultural resources/inputs will be used more efficiently, resulting in greater agricultural sustainability. It is difficult to meet food demands with the planet's limited resources. To meet this challenge, a number of cutting-edge agricultural technologies are being implemented. Precision agriculture (PA) is a collection of sensors and field sensing systems that aid in the monitoring of crop conditions at various stages of development. PA entails gathering and analysing vast amounts of crop health information. PA enables a farmer to understand exactly what parameters are needed for a healthy crop, where they are needed, and in what quantity at any given time. Initially, it is primarily used to adjust fertiliser charges surrounding a subject's high, medium, or low-producing zones. Precision farming allows for greater accuracy and efficiency in all agricultural operations, from sowing, fertilising, irrigation and protection to harvesting, using precision tools such as GPS, GIS, VRT and remote sensing. PF has enormous environmental prospects, such as reduced nitrate leaching, greenhouse gas emissions, and agrochemicals such as fertiliser, herbicides, pesticides, and so on, which may aid in achieving sustainable agriculture in the future.

Figure 1.4: Precision farming

E. Use of GPS, GIS and remote sensing

GPS is a global radio navigation system made up of 24 satellites and ground stations that gives geographic accuracy to farm practices by allowing farmers to identify each field site. Crop rotations in a region can be tracked year to year using a set of maps with each field labelled with the crop planted in each season. GPS is required for yield mapping, crop scouting, and variable input applications to achieve a precision of 1-3 m. Field border and topographical maps can also be created using GPS.

GIS is a kind of computerized map. It provides information on field topography, soil types, surface drainage, subsurface drainage, soil testing, irrigation, chemical application rates and crop yield. "Brain of PF". Land ownership, crop cover, soil type, topography, N,P,K, and other nutrient levels, soil moisture, pH and other base maps are all included in a precision agriculture GIS. Rotations, tillage, nutrient and pesticide applications, yields, and other information are also maintained. Fertility, weed, and pest intensity maps, as well as prescription maps that depict appropriate application rates of farm chemicals at various field areas, are all created using GIS.

Remote sensing has been used in soil mapping, terrain analysis, crop stress, yield mapping and estimation of soil organic matter. Low-flying aircraft and satellites have come to be a prime source of spatial records because of discounts in each value and time of photograph acquisition and transport. Although the use of remote sensing in agriculture is numerous decades old, improvements in spatial, spectral, and temporal resolutions of recent satellites (e.g., IKONOS) and systems consisting of airborne videography, and the latest traits in new hyper-spectral sensors are more and more turning into use to determine the size, region, and purpose of variant.

F. Site Specific Nutrient Management (SSNM)

Agriculture contributes 70 to 90 percent of nitrous oxide (N_2O) emission mostly from nitrogenous fertilizers. Site Specific Nutrient Management (SSNM) is an approach of supplying supplemental nutrients through right amount, right source, right application rate, right time and right application method on the basis of soil test, crop and climatic condition. SSNM approach does not specifically aim to either reduce or increase fertilizer use; it aims on nutrient application at optimal rate and times to achieve higher nutrient use efficiency and yield as well as low cost and low environmental pollution. SSNM reduces N_2O emissions by reducing total nitrogen application and also avoids nitrogen losses by volatilization, leaching and runoff.

Figure 1.5: Site specific nutrient management

G. Leaf colour chart (LCC)

Leaf Colour Chart has been developed in Japan and it is non-destructive method where green colour intensity of rice leaves is measured to identify the nitrogen requirements by crop plants (Nachimuthu *et al*., 2007). This method is also standardized with chlorophyll meter. In both hybrid and inbred rice cultivation, nitrogen application through LCC has been proved superior to locally recommended dose of nitrogen application by many scientists. It was found that when nitrogen fertilizer is applied as per LCC values, farmers can reduce nitrogen application by 20-30 kg/ha without sacrificing any significant rice yield. N application at LCC value less than 3 in Basmati and at LCC value less than 4 in hybrid rice was found beneficial. Moreover, in LCC based nitrogen management, farmers can also skip basal application of N without any disadvantage in terms of grain yield. Sharma *et al.* (2008) mentioned that 10 to 15 % nitrogen fertilizer can be saved by using LCC.

Figure 1.6: Use of leaf colour chart

H. Laser land leveller

Laser levelling is a process where the land surface is smoothen using laser equipped drag buckets to maintain proper land levelling. Precision land levelling alters the field to create a constant slope of 0 to 0.2%. Large horsepower tractors or soil movers operated with global positioning systems (GPS) or laser-guided instrumentation are used to create constant slope either by cutting or filling the soilsurface. Various studies suggested that laser land levelling can reduce irrigation water application by almost 25% and increase wheat yield by 30% as compared to conventional practices. Similar increase in yield and reduction in irrigation water application can be seen in case of maize cultivation under zero tillage and laser land levelling. Jat *et al.* (2009) also mentioned that the use of RCTs, including zero tillage and laser land levelling reduced irrigation water application by 23 to 45 % while increasing crop yield.

I. Crop rotation and cropping system

Crop rotation or intercropping is another vital RCT. It offers higher diversity in crop field. Legumes grown in rotation with cereals or oilseed crops increases subsequent crop yields and reduces input costs due to nitrogen fixation by legumes and also reduces disease infestation (Pannell, 1995). Greater distribution of soil pores created by root system, better distribution of water and nutrients through the soil profile and increased organic matter content can be seen under crop rotation or intercropping. Selection of proper cropping system such as appropriate crops for the location, number of crops per year and addition of legume or cover crops can increase organic carbon content in soil. Crop rotation also restricts the carryover of pests and diseases from one crop to the next crop through crop residues. Moreover, crop rotation helps the farmers to achieve greater biodiversity at the farm level and get more diversified farm income. It also favours a better use of natural resources.

Figure 1.7: Intercropping

J. Smart Crop and Variety selection

In the village level, farmers traditionally prefer to grow local varieties of the crops resulting poor crop productivity due to heat stress, droughts or floods. Selection of climate smart crop varieties is the best adaptation option. Crops which have wide sowing windows can be sown in broad sowing dates. Different weather calamities like heat or cold wave, flood, cyclone, frost, hail storm, reduce crop yield. So based on the weather forecast and long term research data, we have to select a crop variety which is suitable for a particular region. Sometime introduction of new crops in the cropping system can also be a great option against climate change. For example, introduction of a short duration summer legume like moong bean in rice-wheat cropping system as catch crop after harvesting of wheat crop maintain the soil quality by adding organic matter and also reduces the N_2O emission from the field as the residual nitrogen can be used by moong bean after wheat.

K. Integrated farming systems

Integrated farming system (IFS) has enormous potential to make farmer climate resilient through the cultivation of different crops on the same land and using farm resources sustainably. IFS is often less risky as it manage the farm more efficiently and reduce the dependency on input. Proper recycling of by-products, crop residues, weeds and all other farm wastes in combination with conservation of farm resource can reduce chemical load in the form of inorganic fertiliser by 36 percent (Gangwar *et al.*, 2014).

L. Rainwater harvesting

According to World Bank, India having geographical area of 3.29 million km^2 supports more than 18 percent of world's total population but has only 4.2 percent of fresh water resource. An increase in mean temperature will increase demand for water for evapotranspiration by crops and natural vegetation leading to more rapid depletion of soil moisture. According to the researchers, a 1°C rise in temperature will increase the crop water demand by 2%. So, in the present scenario rain water harvesting shall be a boon for the resource saving adaptation programme. During the rainy season in dry and dryland farming areas, rain water harvesting systems such as water tank, dug well, percolation tank, farm pond etc. are used to collect and catch rain water. This water is used by the people during the dry season when there is a shortage of water. Micro catchments (around 1000 sq. m), small farm reservoirs (1000 to 5,00,000 cu. m), rooftop systems, water spreaders, inter row harvesting, runoff farming, small pits, small runoff basins, run-off strips, macro-catchments and flood water system may reduce the water stress as 1-2 supplemental irrigations can be provided to the crops.

Figure 1.8: Rainwater harvesting

M. In-situ water harvesting

In-situ water harvesting technique is the process of collection and storage of rainwater in the soil by intercepting the run off. Prolonged retention of rainwater on the soil surface increases the amount of infiltrated water into the soil. Some effective methods of in-situ water harvesting are off season tillage, deep summer ploughing, broad bed and furrow system, ridges and furrows, compartmental bund making, contour farming etc.

Figure 1.9: In-situ water harvesting

N. Off season tillage or summer ploughing

Off season tillage or deep summer ploughing helps to maintain higher soil moisture resulting to secure optimum plant growth. Deep ploughing is done either by country plough or by blade harrow depending on the soil types during non-crop season. Deep ploughing is done to open the soil pores for more water intake from rainfall or irrigation. Water use efficiency (WUE) of crops can be improved by selecting appropriate crops and cropping systems based on availability of water supplies and rate of seasonal Evapotranspiration (ET). The later can be achieved by adopting of proper irrigation method, irrigation scheduling, tillage practices, mulching and fertilizer application.

Figure 1.10: Summer ploughing

O. Direct Seeded Rice (DSR)

Direct Seeded Rice (DSR), also known as the 'broadcasting seed technique,' is a water-saving method of rice cultivation. In this method, seeds are broadcasted or drilled into the fields directly. In contrast to the traditional water-intensive method of transplanting rice seedlings from a nursery to waterlogged fields, this method saves groundwater. The shortage of labour and water scarcity and decrease soil fertility are the main reason that are increasing farmers' interest in shifting from transplanting to direct seeded rice (DSR). In North-west India, about 35 to 57% of water savings have been reported in different research experiments on DSR (Singh *et al.*, 2002). Direct-seeded rice grown on raised beds can decrease water use by 12–60% compared to flooded traditional transplanted rice.

Figure 1.11: Direct Seeded Rice

P. System of Rice Intensification (SRI)

SRI was evolved in the 1980s in Madagascar. This rice cultivation system increases paddy yields up to 15 t/ha even on infertile soils with less irrigation water and without any external additional inputs (Stoop *et al*., 2002). In this system transplanting of young seedlings is done in a square pattern with wide spacing, use organic fertilizers and hand weeding, and keeping the field soil moist during vegetative phase. Significant changes occur in plant phenotypic structure, function, grain yield and other yield components under SRI method. SRI increases yields by 50–100% with using only half of the water required under conventional method (Uphoff *et al*., 2010), and no additional external inputs are needed.

Q. Contour farming and strip cropping

Farming on the contour reduces runoff, increases the water infiltration rate and reduces the soil erosion. Contour farming can reduce soil erosion by 50% compared to farming up and down hills. The hill areas where sheet and rill erosion is a major problem on sloping land, the strips are laid out across the hill slope and the areas where wind erosion is a major problem, the strips are laid out closely at the wind direction. Thus strip cropping helps to reduce soil erosion, reduce dust emission into the atmosphere, improve water quality, improve soil health, improve visual quality of the landscape and improve wildlife habitat.

Figure 1.12: Counter farming

1.5 Machinery development and CA

The real success of conservation agriculture in South Asian countries started with the development of modern machineries. At present four machines are available for seeding direct-seeded rice and wheat. Double disc coulter, rotary disk drill (RDD), star wheel, punch planter, and turbo happy seeder are some example of machine that are widely used on CA. Though on CA the major priority is given on RCTs such as zero or reduced tillage, crop residue management, raised bed planting, direct-seeded rice and other RCTs, new tools and techniques such as SPAD, LCC, Green Seeker etc are also now getting importance.

1.6 Constraints

Conservation agriculture offers solutions to break productivity barriers and sustain natural resources and environmental health. Despite having several benefits, in developing countries the adoption of CA by the farmers is still in the initial stage. CA technologies are basically machine driven, herbicide driven and knowledge driven. Therefore, vastly improved expertise and resources are required for adoption in large areas. For adoption of CA in wider space, there is an urgent need for researchers and farmers to change their mindset and explore these opportunities in a site and situation specific manner for local adaptation.

Several factors limit the adoption of CA by resource-poor farmers. The current major barriers that restrict the spread of CA system are (i) competing use of crop residues in rainfed areas, (ii) weed management strategies, particularly for perennial species, (iii) localized insect and disease infestation and (iv) livelihood of lower crop productivity if site-specific component technologies are not adopted. There are several other factors that restricts the adoption of CA technologies by the farmers in India (Table 1.2)

Table 1.2: The nature of constraints towards adoption of CA technologies in India

Technical	Non-availability of quality drill
	Lack of proper manufacturers of machines
	Lack of training for the farmers
Extension	Lack of support from state government
	Lack of extension literatures
	Lack of attention by mass media
	Lack of proper knowledge about CA
	Inadequate extension facility
	Lack of cooperation from fellow farmers
Financial	Lack of credit facilities
	Lack of money to buy new machines and inputs
	No subsidy on inputs
	High cost of machineries

Source: Meena and Singh (2013)

References

Bhattacharyya R, Pandey SC, Bisht JK, Bhatt JC, Gupta HS, Tuti MD, Mahanta D, Mina BL, Singh RD, Chandra S, Srivastva AK and Kundu S (2013a). Tillage and irrigation effects on soil aggregation and carbon pools in the Indian subHimalayas. Agron. J. 105: 101-112.

Foresight (2011). The Future of Food and Farming. Final Project Report. The Government Office for Science, London.

Gathala MK, Ladha JK, Saharawat YS, Kumar V, Kumar V and Sharma PK (2011). Effect of tillage and crop establishment methods on physical properties of a medium-textured soil under a seven-year rice–wheat rotation. Soil Sci. Soc. Amer. J. 75: 1851-1862.

Jat ML, Saharawat YS and Gupta R (2011). Conservation agriculture in cereal systems of south Asia: Nutrient management perspectives. Kar. J. Agric. Sci. 24: 100-105.

Ladha JK, Kumar V, Alam MM, Sharma S, Gathala MK., Chandna P, Saharawat YS and Balasubramanian V (2009). Integrating crop and resource management technologies for enhanced productivity, profiability and sustainability of the rice–wheat system in South Asia. In: Integrated Crop and Resource Management in the Rice–Wheat System of South Asia. IRRI, Los Ba˜nos, the Philippines, pp. 69-108.

Pandey S, Byerlee D, Dawe D, Dobermann A, Mohanty S, Rozelle S and Hardy B (2010). Rice in the Global Economy.

Saharawat YS, Singh B, Malik RK, Ladha JK, Gathala M, Jat ML and Kumar V (2010). Evaluation of alternative tillage and crop establishment methods in a rice–wheat rotation in North Western IGP. Field Crops Res. 116: 260-267.

2

Resource Conservation Technologies for Soil Health Management

2.1 Introduction

Present day agriculture is facing more challenges like negative impacts of climate change, depletion of natural resources, increase of input costs etc. The major causes for steady decline in agricultural sustainability include intensive cultivation with conventional tillage practices, reduction of soil organic matter, soil loss by water and wind erosion, reduction in infiltration rate, formation of soil crust and continuous practice of mono-cropping. Over the years, soil and water resources are continuously overexploited leading to reduction in use efficiency of irrigation and energy. Rice-wheat cropping system is mostly practiced by the farmers in Indo-Gangetic Plain region of India with unsustainable, conventional land preparation and sowing practices which not only disturb the soil environment but also lead to atmospheric pollution. During green revolution, food grain production enhanced mainly from increase in rice-wheat area as well as the system productivity. Continuation of the same rice-wheat cropping system without proper crop rotation over the last 50 years has created many problems which have restricted the ability of natural resources to produce higher yields to match with future food grain requirement of the country. The deteriorating production and sustainability of the rice-wheat systems are evident from either stagnation or decline in soil physical environment, excessive mining of essential plant nutrients from soil. Use of water to irrigate rice and wheat has surpassed the natural ability of the ecosystems to replenish itself. In recent years growing declaration in total factor productivity and deterioration of soil health under cerealbased cropping system has necessitated for diversification of existing cropping systems with conserving the natural resources efficiently. Therefore, simple change or management of cultivation practices through eliminating or modifying the unsustainable conventional agricultural practice like conventional tillage, removal of all organic material, monocropping are needed for gain in productivity in near future while sustaining the natural resource base.

The term soil health is used in relation to sustainability in agriculture. Soil health includes physical, chemical and biological properties of soil and their integration. Soil quality is a term that is used to describe physical attributes of a soil. These characteristics are simple traits like soil colour to more complex properties like fertility, erodibility and compatibility. Soil health is manifested through vital soil properties or through key soil functions. A healthy soil is supposed to have values of key parameters or key soil functions in the desirable and optimum range. Thus, the concept of soil health quality is based purely on soil properties and it sustains plants, animals and humans, maintains the quality of water. Intensive agricultural practices with tillage operation lead to gradual deterioration in soil physical, chemical and biological properties and consequently reduction in crop productivity. Thus, practice of conventional tillage operation over few decades has been questioned because of the soil erosion from cultivated land after tillage. It is a well-known that injudicious use of farm inputs and over exploitation of natural resources mainly land and water had led to the following major soil health issues: ground water depletion, loss of soil organic carbon, soil salinization, soil compaction and increase in the runoff rate resulting more soil erosion. Deterioration of soil physical and chemical properties, pollution of ground water in some locations due to over application of nitrogenous fertilizers, harmful effect on the environment through crop residue burning and pesticide use have been emphasized (Gupta *et al.*, 2003). In eastern IGP region, rice-wheat system is mainly practiced by operating intensive labour and less mechanization. Crop residue removal/ burning in India is a serious problem, leading to widespread problems of secondary and micronutrient deficiency, explaining why we have the "yield stagnation" scenario presented so often. The current challenge is addressing the issue of small land holdings and poverty amongst these farmers. Unfortunately, poverty has made crop biomass an attractive fuel for farm homes, and an additional source of income. As farm size increases, alternative fuels (gas/ electricity) will become more popular and crop biomass will no longer be in demand.

In the era of faster economic development, maintaining and protecting soil health under intensive land use system is a major challenge to maintain sustainability of agriculture resources especially in the developing countries like India. The basic assessment of soil health is necessary to evaluate the changing trends following different land use and management interventions. Soil health is adversely affected due to excessive fertilizer use and soil pollution, soil loss and nutrient imbalance in soil processes (Swaminathan, 2006). The soil health concept has evolved with an emphasis on understanding the soil processes. Soil physicochemical and biological properties are sensitive to management practices can be used as indicators in measuring the soil health. Lal (1994 a, b) suggested that soil degradation caused by improper use of land and soil particularly its management should be quantified by measuring management induced changes in soil properties and their impacts on actual and potential productivity. Establishing and understanding the cause–effect relationship between soil properties and processes on the one hand and crop productivity and environmental moderating functions on the other is crucial to enhancing soil productivity, restoring degraded lands and improving environmental quality. Under such scenario, resource conservation technologies (RCTs) assume great significance.

2.2 Resource conservation technologies for soil health

Soil affects not only the production, but also the management of other natural resources such as water, which are associated with soil. Soil structure is strongly correlated with the content of organic matter which gives life to the soil. Organic matter stabilizes soil aggregates, provides feed to micro-organisms and helps to hold soil water. Due to use of intensive conventional tillage-based agriculture systems, the organic matter content of soil is steadily decreasing, leading to decrease in crop productivity, followed by the visible signs of soil degradation and finally desertification. We are aware that most of the biological, chemical and physical properties which are attribute to improved soil health, come from the maintenance or improvement of soil organic matter content of the soil (Hati *et al.*, 2008). Soil organic matter not only improves soil health, water infiltration rate, water storage capacity, CEC and microbial activity but also the availability of macro, secondary and micronutrients. There is a deterioration of soil health parameters, such as soil organic C, soil biodiversity, soil physical properties and soil fertility. There is a gap of more than 10–12 million tonnes of NPK between the nutrients depleted by crops and nutrients added through fertilizer. Farmers in the developing countries use low inputs because of their socio-economic constraints and serious problems of drainage during heavy rains and rainwater management in drought situations. In the Indo-Gangetic Plains, decrease in yield of high yielding varieties due to increase doses of fertilizers and degradation of soil health resulting from overexploitation of natural resource base can be seen. Conservation of land resources for agriculture is necessary to stop the extent of land degradation across the India. In India, Jammu & Kashmir and Nagaland states have maximum share of their land (94%) under degradation because their large areas are under mountains, cold deserts and other such degraded lands. The extent of land degradation is also a great concern in some agriculturally prominent states like Uttar Pradesh (63%). Madhya Pradesh (50%) and Karnataka (46%). Natural farming through conservation agriculture will help to check land degradation. Long-term trials in South Asia clearly showed that well managed high yield cropping systems significantly improved the soil organic matter and related soil biological, chemical and physical properties by returning crop residues to the field (Singh *et al.*, 2016). Further work with conservation tillage and nutrient management mentioned that the system not only improved soil organic matter, but also supported higher crop yields and reduced associated GHG emissions per tonne of grain production (Sapkota *et al.*, 2014). Resource conservation technologies for soil health management are mentioned below.

A. Minimum soil disturbance through conservation tillage

Conservation tillage practice refers to tillage system where at least 30% of the soil surface is covered with crop residues. After sowing or planting to reduce water erosion or where wind erosion which is the primary concern maintain, at least 1000 kg/ha of crop residues should be present on the soil surface during the critical wind erosion period. Conservation tillage (CT) is described as tillage practices specifically intended to reduce soil disturbance during seedbed preparation with the objective of improving soil organic matter, structure and stability, and conserving soil moisture. It is intended to conserve soil, water, energy, and protect water and environmental quality. Conservation tillage systems involve a variety of techniques, including no-till, minimum till, ridge till, chisel plough etc. This tillage technique can be key component of a sustainable agricultural system, as it decreases losses due to soil erosion, enhances input use efficiency, increases soil organic matter build up and reduces the cost of production and environmental degradation. It is important to remember that any action that is done to decrease soil erosion also reduces the need to add fertilizers and water to the soil with the fact that most of the organic matter contents present in top soil. Conservation tillage reduces water pollution and burning of fossil fuel energy resulting decrease in CO_2 emission and cost of production compared to other conventional tillage practices.

B. No-tillage or zero tillage

In no-tillage system, soil is not disturbed from planting to harvest except sowing and fertilizer application. Weed control is done by using herbicides only. No-till farming, a conservation tillage technique synonymously called as zero tillage or direct sowing, is a way of growing crops from year to year without disturbing the soil through tillage operations. No-tillage operation has been developed to maintain soil health for sustainable agriculture and economic alternative. It contributes to carbon sequestration through storage of soil organic matter in the soil (Bayer *et al.*, 2006). Under conventional tillage operation soil layers turned up and down, air mixes in and soil microbial activity increases over baseline level resulting faster oxidation and breaking of organic matter and carbon is lost from the soil into the atmosphere. No-till farming also reduces nitrous oxide (GHG that stays in the atmosphere for 120 years) emission by 40–70% depending on the crop rotation (Omonode *et al.*, 2011).

C. Reduced or minimal tillage

Before sowing little disturbance of soil for loosen the compact soil and seedbed preparation refers to reduced or minimal tillage. Weed control is done by using herbicides or some secondary tillage implements. Minimum tillage is also a kind of conservation tillage practices like strip tillage with minimum soil disturbance for a successful crop production. It is a tillage operation that does not turn the soil surface, reduces the evaporation loss and enhances the water holding capacity of the soil. It is contrary to intensive tillage, which changes the soil structure by plough.

D. Ridge tillage or raised bed planting

It is one type of conservation tillage technique where equipment is mainly used to move soil so the crop row is slightly elevated. This system is usually used on soils that drain poorly or warm slowly in the spring. The rows are maintained each season in the same place and the top of the ridge is levelled off and planting is done. Planting of crop plants on these ridges provides warm condition for early plant growth. In most of the times, the crop residues on the soil surface are not disturbed. Furrow irrigated raised bed (FIRB) is a one type of ridge tillage technique slowly popularizing in Indo-Gangetic Plan (IGP) for maize based cropping system which is a common practice in western countries (Gupta *et al.*, 2002).

E. Mulch tillage

Tillage is done only for sowing the crop and equipment doesn't use to bury the crop residues in the field. Weed control is done by using herbicides or some secondary tillage. Mulch tillage also called stubble mulch tillage is the practice of managing the amount, orientation and distribution of high percentage of plant residues on the soil surface as cover throughout the year. Mulch tillage technique prevents the soil erosion (more effective for wind erosion), reduces soil compaction, improves soil tilth and air quality, organic matter is added to soil, improve water quality and save time and money. Hobbs *et al.*, (2008) clearly compared the conventional tillage practice and conservation tillage practice (Table 2.1).

Table 2.1: A comparison between traditional tillage (TT) and conservation tillage (CT)

Issues	Traditional tillage (TT)	Conservation tillage (CT)
Practice	Disturbs the soil surface and leaves a bare surface	Reduces the soil disturbance and keeps the soil covered
Erosion	Wind and soil erosion is maximum	Both the erosions reduced significantly
Soil physical health	Reduced	Significantly improved
Compaction	Reduces compaction by destroying biological pores	Reduced tillage does not reduce soil compaction
Soil biological health	The lowest due to frequent disturbance	Moderately better soil biological health
Water infiltration	Lowest due to soil pores clogged	Water infiltration rate is good
Soil organic matter	Oxidizes soil organic matter and causes its loss	Soil organic build up possible in the surface layers
Weeds	Controls weeds and also causes more weed seeds to germinate	Reduced tillage controls weeds
Soil temperature	Soil surface temperature is more variable	Intermediate in variability
Diesel use and cost	Diesel use: High	Intermediate
Production costs	Highest	Intermediate
Yield	Can be lower where planting is delayed	Yields same as TT

Source: Hobbs *et al.* (2008)

F. Conservation Tillage and Soil Health

In conventional tillage operations bare soil is exposed to air and water which are the major causes of soil loss. Under conservation tillage, the crop residues buffer the raindrop's energy, so water has less erosive force when it touches the soil. This protection by crop residue along with the rougher surface provided by the residue facilitates infiltration and decreases runoff that carries soil and nutrients with it. In addition macro-pores, which are the major routes for water movement through the soil profile, get disturbed in tillage operation but remain intact under conservation tillage practices. Conservation tillage practices also enhance water infiltration, decrease runoff and conserve water. Another important benefit of no-till production system is higher soil organic matter concentration near the soil surface (Dick, 1983). It was reported by Hargrove and Fyre (1987) that organic matter accumulation in soil was higher under no-till management and accumulation rate depended largely on the amount of organic carbon returned to the soil (Table 2.2).

Table 2.2: Influence of cropping sequences and different types of tillage practices on soil organic C and N concentrations in the surface soil

Crop sequence	Tillage practice	Organic C (%)	Organic N (%)
Wheat-soybean	Conventional	1.4	0.12
Wheat-soybean	No-till	1.6	0.15
Clover-soybean	Conventional	1.9	0.14
Clover-soybean	No-till	2.2	0.17

Table 2.3: Crop residue and nutrient returned from rice straw

Tillage	Crop residue (t/ha)		Residue (kg/ha)			
	Above ground	Root mass	C	N	P_2O_5	K_2O
Conventional till	-	6.86	3090	36.4	17.28	128
No-till	3.13	6.87	4340	53.0	25.19	187

Source: Pasricha (2014)

Table 2.4: Grain yields, harvest index and WUE in wheat under two different tillage treatments

Tillage	Grain yield (t/ ha)	Harvest index (%)	Water use efficiency (kg grain/ha/ mm)
Conventional till	5.95	44.0	7.02
No-till	6.07	46.2	8.14

Efficient use of irrigation water through conservation agricultural practices, crop residues management and zero tillage are most appropriate ways to restore the soil fertility (Pasricha, 2014). On long-term basis, such practices help in sequestering organic C in the soil and recycle part of the nutrients. These practices thus help in maintaining the quality of environment by preventing the burning of the crop residues in the field.

G. Maintaining soil cover

Maintaining a permanent or semi-permanent cover on the soil surface with any organic material is necessary to protect the soil from sun, rain and wind, and to feed soil microflora. Soil cover can be maintained by cultivation of cover crops, use of mulching and crop residue management.

H. Cover crops

Cover crops should be multi-purpose crops and should be managed appropriately. For example, *Crotalaria juncea* L. which is used as a cover crop and also a good source of fibre is very effective for nematode control and fertility improvement through N fixation. Good ground cover by crop canopy of cover crops especially legumes protect the soil from erosion. Cover crops also add organic matter by litters. Besides conserving soil and moisture, the cover crops hold those soluble nutrients, which are lost by leaching. Among the legumes cowpea has been found to produce maximum canopy in rainy season followed by horsegram, greengram, blackgram and sesbania. These crops give early and dense ground cover up to the extent of 85% which generally coincides with peak rate of runoff. The permanent soil cover through dense crops, mulch or green manure cover crops complements the impacts of zero tillage by supplying substrate for soil organic matter build up and for the soil which is facilitated by not disturbing the soil. The mulch protects the soil surface, reduces evaporation loss, avoids crusting and suppresses weed growth. Problems faced in direct seeding or zero tillage can be reduced thus. Adoption of zero tillage and direct seeding also facilitates the management of residues which are often created a problem in conventional systems.

I. Mulching

Mulching is a layer of any material that covers the soil surface. Mulching can either be organic or inorganic. It prevents soil erosion, checks surface runoff and improves soil health by adding organic matter. Use of mulching management in maize-wheat cropping system revealed that use of polyethyl mulch recorded the highest grain yield followed by use of straw mulch (Sharma *et al.*, 2009).

Table 2.5: Effect of moisture conservation practices on total water use, WUE, yield and net return (mean of 2005–2016 and 2006–2007)

Moisture conservation practices	Total water use (mm)	WUE (kg seed/ha mm)	Seed yield (kg/ha)	Net return (Rs/ha)
Control	187.0	5.05	9.48	13,262
Dust mulch (created by weeding and hoeing at 20 and 30 DAS)	184.7	6.41	11.85	17.142
Organic residue mulch (4 t/ha)	180.6	7.60	13.69	21,138

Source: Yadav *et al.* (2011)

Table 2.6: Effect of mulch rate on water transmission properties

Mulch rate (Mg/ha)	Soil water absorptivity	Soil water transmissivity	Saturated hydraulic conductivity (cm/h)
0	0.32	5.56	30a
2	0.57	7.81	45a
4	0.67	7.50	70b
6	0.84	10.21	132c
12	1.05	15.36	129c

Source: Lal (1987a, b)

J. Crop residue management

Continuous decrease of soil organic matter in the soil due to limited return of organic matter is identified as the major reason for unsustainability of the system. Burning of residues after harvesting of the crops without in-situ recycling not only leads to loss of N, P, K and S in a considerable amount but also contributes to the global NO_2 and CO_2 budget and destruction of beneficial micro-organisms present in the soil profile.

K. Crop rotation

Crop rotation is one of the important agricultural management systems with ancient origin. The rotation of different types of crops having different rooting patterns in combination with zero-tillage or minimal soil disturbance promotes a network for root growth and macro-pores in the soil. This also helps in infiltration of water to the deeper soil depths. Crop rotation increases microbial diversity in the field and as a result the risk of pest and disease is reduced. Lowland rice grown in rotation with upland crops including legumes has the potential to sustain the system productivity and to reduce problems related to the soil health. Moreover, such need-based crop diversification fulfils the needs of farmers through cereal, pulse, oilseed and vegetable crops, and maintain their income by withstanding weather hazards, reducing use of chemical fertilizers and pesticides, conserving natural resources, ensuring a balanced food supply, ensuring environmental safety, controlling price fluctuation and generating employments for farmers. In the time of shrinking natural resources, enhancing the water and energy use efficiency is important to maintain suitability of any cropping system.

L. Agroforestry and soil health

In agroforestry crops, trees and animals are cultivated together in a long-term eco-friendly system where more emphasis is given on perennial multi-purpose trees that are planted once but farmers get yield benefits over a long period. Agroforestry is recommended by scientists and planners for sustainable agricultural production, natural resource conservation, soil erosion management and recently for mitigation of climate change by the way of sequestering carbon in both standing biomass and soil.

M. Natural resource conservation

Agroforestry helps to conserve soil, water and biodiversity. Reduction in soil erosion, check in runoff and increase in soil fertility due to litter falls are observed as compared to conventional agriculture practices or fallow lands in different agro-climatic regions. Some regional examples of agroforestry system in India are mentioned in Table 2.7.

Table 2.7: Regional examples of effect of agroforestry on soil health

Region	Challenges	Changes observed due to agroforestry	Reference
Western Himalayas	Minimizing soil and water erosion in agro-ecosystems on steep slopes	Contour tree rows (hedge rows) reduced runoff and soil loss by 40 and 48%, respectively	Narain *et al.* (1997)
Sikkim Himalayas	Enhancing litter production and soil nutrient dynamics	N-fixing trees increased N and P cycling through enhanced litter production, resulted in greater availability of N and P, maintained soil organic matter, higher N mineralization	Sharma *et al.* (1996a, b)
Himalayas	Rehabilitation of degraded, abandoned agricultural and mine spoiled sites	Enhanced biomass accumulation (3.9 t/ha), soil physico-chemical properties, C-sequestration	Maikhuri *et al.* (2000)
Kurukshetra Himalayas	Problem soil improvement	Increased soil microbial and tree biomass, available N	Kaur *et al.* (2002)
Western India (Karnal)	Soil fertility improvement of moderately alkaline soils	Soil microbial biomass C increased (109.12–96.14 gg−1 soil) under tree plantation; soil carbon increased by 11–52% due to integration of trees and crops	Kaur *et al.* (2000)
Central India	Soil improvement	Soil organic C, mineral N and P increased	Pandey *et al.* (2000)

N. Nutrient recycling

Agroforestry is known for more efficient nutrient recycling than cropping alone. In forest ecosystem, nutrient inputs from biological nitrogen fixation, litter fall, decay of plant residue and weathering of primary soil minerals balances nutrient losses due to plant removal, denitrification, leaching and runoff. However, there is a net removal of nutrients from the soil which can result in negative balances of nutrients if the nutrients are not provided through external inputs like fertilizers and manures. Planting of trees in agricultural land-scape can provide additional nitrogen in soil through biological nitrogen fixation (BNF), deep nutrients capturing capacity and nutrients recycling ability. The presence of active root nodules in legumes indicates that BNF can supply considerable nitrogen inputs to crops. The nitrogen non-fixing trees such as acassia can accumulate more nitrogen in their leaves than nitrogen-fixing legumes, probably because of their large root volumes and nutrient capturing capabilities (Garrity and Mercado, 1994). This nitrogen is again added to the soil as green leaf manure. *Sesbania* and *Gliricidia* are cultivated in the field for their nitrogen fixation ability and green manuring potential. In some plants like *Leucaena*, the above ground litter contain high N (about 256 kg/ha) with a yearly release of almost 208 kg/ha (Sandhu *et al.*, 1990). Nutrients captured by tree roots at the soil depths where usually crop roots cannot reach is considered as an additional nutrient input in agroforestry because these soil nutrients are otherwise leached out. Tree litter decomposition is also an input of nutrients to the soil.

Agroforestry can provide the required nutrients especially nitrogen to the crops. Generally about 20% of the nitrogen released from pruned plant material or litter fall is up taken by the current crop plantss and remaining portion is incorporated into soil as organic matter (Haggar *et al.*, 1993). As N and P are slow release nutrients from SOM, organic matters have a greater residual effect than inorganic fertilizers on soil fertility. The rate of decomposition mainly depends on climatic variables, edaphic and resource quality (Yadav *et al.*, 2008). Organic matters with high N and low lignin and polyphenol content release nutrients rapidly while those with low N and high polyphenol and lignin content release nutrients very slowly or even immobilize (Palm, 1995). Maharudrappa *et al.* (1999) mentioned that incubation of litter from different multi-purpose trees enhanced nutrient availability and available N increased after 60 days of incubation. This indicates temporary immobilization of nutrients by soil microorganisms. Legumes constitute the cheapest source of protein to man and animals, enhance soil quality by adding organic matter and improve soil structure and water infiltration (Palsaniya and Ahlawat, 2007). They act as nurse crops and supply nitrogen to the associated crops and contribute to the conservation of energy by reducing the need for N fertilizer (Palsaniya and Ahlawat, 2009). Unlike N, the release of P to the available pool from litter was recorded at 30 days of incubation. Mineralization of organic matter also played a major role in release of P for plant growth. The decomposition of litter with narrow C:P ratio increased the amount of available phosphorus in the soil. Incubation of litter also increased availability of K in the soil. The release of K into the soil mostly depends on quality and quantity of litter. The increase of available K in soil indicates that K is not strongly bound in organic matters unlike that of N and P.

2.3 Limitations

A change in the mentality of farmers, technicians, extension workers and researchers away from conventional soil degrading tillage operations towards sustainable production systems like no-tillage or minimum-tillage is necessary. Convincing the farmers that successful cultivation is possible even with reduced tillage or without tillage is a major barrier in promoting CRA on a large scale. It is also difficult to convince the farmer about the potential benefits of CRA beyond its potential to reduce production cost, mainly by reducing tillage operations. The following are a few important constraints which restrict the broad scale adoption of CRA for soil management.

- **Lack of appropriate seeders especially for small and medium scale farmers:** Although significant efforts have been made in developing and promoting machinery for seeding wheat in no till systems, successful adoption will call for accelerated effort in developing, standardizing and promoting quality machinery aimed at a range of crop and cropping sequences. These would include the development of permanent bed and furrow planting systems and harvest operations to manage crop residues.
- **The widespread use of crop residues for livestock feed and fuel:** Under rainfed situation farmers face scarcity of crop residues due to less biomass production by crops. There is also a competition between CA practices and livestock feeding for crop residues. It is a major constraint for adoption of CA under rainfed situations.
- **Burning of crop residues:** For timely sowing of the next crop and lacking of machinery for sowing under CA systems, farmers prefer burning of crop residues to sow the crop in time. This has become very common in rice-wheat system in north India.
- **Lack of knowledge about the potential of CA to agriculture leaders, extension agents and farmers:** Whole cultivation practices in conservation agriculture including planting, water and nutrient management, diseases and pest control, harvesting etc., need to be introduced among agriculture leaders, extension agents and farmers.
- **Skilled and scientific manpower:** Managing conservation agriculture systems need killed and scientific manpower. Strengthened knowledge and information sharing mechanisms are also needed.

References

Hobbs PR, Sayre K, Gupta R (2008). The role of conservation agriculture in sustainable agriculture. Philos Trans R Soc B 363:543–545.

Kaur B, Gupta SR, Singh G (2000). Soil carbon, microbial activity and nitrogen availability in agroforestry systems on moderately alkaline soils in northern India. Appl Soil Ecol 15:283–294.

Kaur B, Gupta SR, Singh G (2002). Carbon storage and N cycling in silvipastoral systems on sodic soils in north-west India. Agrfor Syst 54:21–29.

Lal R (1987a). Farming systems. In: Lal R (ed) Tropical ecology and physical edaphology. Wiley, New York, pp 618–685.

Lal R (1987b) Tropical ecology and physical edaphology. Wiley, New York, pp 618–685.

Pasricha NS (2014). Success story of resource conservation technology in Indo-Gangetic Plains. In: Ghosh PK, Kumar N, Venkatesh MS, Hazra KK, Nadarajan N (eds) Resource conservation technology in pulses. Scientific Publishers, Jodhpur, pp 32–42.

Sharma R, Sharma E, Purohit AN (1996a). Cardamom, mandarin and nitrogen fixing trees in agro-forestry systems in India's Himalayan region. I. Litterfall and decomposition. Agrfor Syst 35:239–253.

Sharma R, Sharma E, Purohit AN (1996b). Cardamom, mandarin and nitrogen fixing trees in agroforestry systems in India's Himalayan region. II. Soil nutrient dynamics. Agrfor Syst 35:255–268.

Yadav PN, Uttam SK, Singh RP, Kumar K (2011). Effect of fertilizer and moisture conservation on productivity, economics and water use efficiency of rainfed Indian mustard (*Brassica juncea*). Curr Adv Agric Sci 3(2):108–111.

3

Resource Conserving Technologies in Rice Cultivation

3.1 Introduction

Rice is the main staple food for most of the world's population and plays a key role to maintain food security of many countries. More than 90% of the rice production globally comes from Asia and also more than 90% of rice consumption is done in Asia (IRRI, 1997). In Asia, more than two billion people get almost 60-70 percent of their daily energy requirement from rice and its products (Yogeswari and Porpavai, 2018). As for India, since independence the country has witnessed a remarkable progress in rice production. In India, rice cultivation occupies an area of 43.79 million ha with an average 112.91 million tones production and 2.58 tonnes/ha productivity (Anonymous, 2018).

The increasing population growth at the rate of 1.4% (Duttarganvi *et al.*, 2016) is creating burden on land and other natural resources base of the country, which damage the ecological base of agriculture system. Rice crop requires more water than other cereal or pulse crops due to its low water use efficiency. The increased rice cultivation over the years has accelerated the removal of soil nutrient four times during the last four decades exerted fourfold pressure on soil resources. Furthermore denitrification, volatilization losses and nutrients leaching was reported than earlier grown rice ecosystem. This leads to lower nutrient use efficiency and more nutrient loss. For rice production cost of cultivation is high due to excess use of inputs and increasing the energy demand needed of tillage operation, fertilizers requirement, number of irrigation needed for rice production. These problems can be addressed through adoption of resource conservation technologies.

3.2 Traditional methods of rice cultivation

A. Puddled Transplanted Rice (TPR)

In India, rice is the main staple food and a source of livelihood for about 120-150 million rural households (Nivrutti, 2014). Irrigated rice occupies almost 50 % area and contributes nearly 70% to total rice production of India with an average yield of 3.1 t/ha. To produce 1 kg rice grain, farmers need to provide 3-5 times more water in the field than any other cereals (Kumar *et al.*, 2013). Transplanted rice is predominantly cultivated in the North-Western Indo-Gangetic Plains (IGP). Transplanting requires at least 25 ha-cm of water for puddling operation, which creates a dense clay layer in the sub-soil to prevent seepage losses. In traditional rice cultivation, rice seedlings are prepared in the nursery and then rice seedlings are transplanted in puddled field into standing water. The amount of water required for puddling of the field varies from 100 to 544 mm (Sudhir-Yadav *et al.*, 2011). Singh *et al.*, (2014) mentioned that in case of transplanted rice about 20 – 40% of the total available water was required for growing of rice crop and subsequently create difficulties in seed bed preparation for succeeding wheat crop in rotation. It, therefore, becomes imperative to identify alternative establishment method to puddling especially in those regions where water is becoming scarce, and an upland crop is grown after rice.

Figure 3.1: Transplanting of rice

3.3 Resource conservation technologies in rice

RTCs are any practices which improves input or resource use efficiency. RCTs protect top soil, improve water holding capacity and increase rice productivity. However, for rice production water is the most essential input as it affects plant growth by influencing physiological processes of crop plants. Availability of fresh water for agriculture is becoming scarce day by day. In Asia, suitable growing condition for rice has declined due to continuously increasing water stress (Tao *et al.*, 2004). Groundwater tables have already dropped below the average level on average by 0.5–0.7 m/year in Rajasthan, Punjab, Haryana, Karnataka Maharashtra, and Northern Gujarat and by 1 m/year in Tamil Nadu. There are various nutrient losses in rice crop through leaching, volatilization and dentrification, which reduces the productivity per unit area. Thus in order to meet the increasing pressure of biotic and abiotic stresses in response to soil degradation, crop productivity and farmers' profits have shown a downward trend. Current efforts have endeavoured to develop and carry RCTs with effective and environmentally friendly crop establishment as compared to the conventional practices. Various resource conservation technologies that make rice production more water-efficient with integrative use of management practices are system of rice intensification (SRI), alternate wetting and drying (AWD), direct seeded rice, aerobic rice, furrow irrigated raised beds etc. Effect of different RCTs on rice yield and water productivity have been presented in Table 3.1.

Table 3.1: Effect of different systems of rice cultivation on grain yield and water productivity of rice

System of rice cultivation	Grain Yield (kg/ha)	Water productivity (kg/m^3)
Transplanted rice	6262	0.37
Alternate wetting drying	5796	0.42
System of rice intensification	6682	0.47
Direct wet seeded rice	5500	0.35
Aerobic rice	3933	0.42

Source: Geethalakshmi *et al.*, 2011

A. Alternate Wetting and Drying (AWD)

Alternate wetting and drying is commonly used as a water-saving rice cultivation practice in many regions of the world. In this system, the soil of the rice field is allowed to dry for a few days within two irrigation schedule. Saving of irrigation water in AWD treatment was about 53– 87 mm (13–16%) compared with submerged regime. Water productivity was significantly higher in the AWD regime than in continuously flood irrigated rice (Belder *et al.*, 2004). Yield penalty was seen under AWD compared with the continuously submerged rice (Bouman and Tuong, 2001). However, the water consumption is still high in AWD since the soils need to be submerged at least during the irrigation period.

B. Direct Seeded Rice (DSR)

In direct seeded rice (DSR) cultivation, raising of nursery for transplantation is excluded. DSR gives the farmer flexibility to take up direct sowing of paddy in case of water shortage or delayed monsoon. Direct sowing can be practiced for cultivating both coarse rice and basmati rice wherever feasible in the North-West IGP region. In DSR, after applying pre-sowing irrigation, first irrigation is applied at 7-10 DAS depending on the type of soil. During active tillering growth phase (30-45 DAS) and reproductive phase optimum moisture should be maintained and irrigation is given at 2-3 days interval to harvest optimum grain yield from DSR. In a study Gathala *et al.* (2011) observed that in dry DSR irrigation could be applied safely when soil hairline cracks appeared. Gupta *et al*., (2002) mentioned that DSR on raised beds decreased water requirement by 12-60%, and increased grain yield by 10% as compared to traditional transplanted rice.

Figure 3.2: Direct seeded rice (DSR)

C. Aerobic rice

Aerobic rice is another new way of rice production system in which specially developed and input-response rice varieties with aerobic adaptation are grown in well-drained, non-puddled, and non-saturated soils without ponded water. The main principal of this approach is to reduce water requirement in rice field like other irrigated upland cereal crops such as wheat or maize. Kadiyala *et al*., (2012) observed that the total amount of water applied in the aerobic rice field including rainfall was 967 and 645 mm compared to 1546 and 1181 mm in flooded transplanting rice during 2009 and 2010 respectively. The result showed almost 37 to 45% water savings in the aerobic rice. Jinsy *et al.*, (2015) also reported that the average water productivity of aerobic rice was 60.7% higher than that of conventional rice cultivation. Reddy *et al.*, (2010) concluded that water productivity was higher under aerobic rice (0.20 to 0.60 kg/m^3) than that under transplanted rice (0.14 to 0.43 kg/m^3). Nevertheless, decline in yield was observed when aerobic rice when continuously grown and the decline were greater in the dry than in the wet season (Peng *et al.*, 2006). In crux, aerobic rice is an attractive option to the traditional rice production system. Yield penalty and yield stability of aerobic rice have to be considered before promoting this water-saving technology.

Figure 3.3: Aerobic rice

D. Furrow irrigated raised beds

Transplanting of rice seedlings on raised beds omits puddling. In this case rice is grown on raised beds and irrigation is applied in furrows between the beds. Naresh *et al.*, (2014) reported that rice cultivation on wide raised beds saved about 15–24% irrigation water and grain yield was decreased by almost 8% than other rice cultivation techniques. Sandhu *et al.* (2012) concluded that transplanting of rice seedlings on slopes of freshly constructed beds saved about 15% of irrigation water as compared to puddled rice fields without any significant reduction of grain yield. Singh *et al.*, (2001) compared grain yield and water use of rice established under transplanting, wet and dry seeding on flatland and on raised beds. The result showed that transplanted rice and dry-seeded rice on flatland and on raised beds reduced total water requirement by almost 35–42% when the soil was kept near saturation, by 47% when soil dried out to 20 kPa moisture tension and by 51% when the soil dried out to 40 kPa moisture tension in the root zone.

Figure 3.4: Furrow irrigated raised beds in rice

E. System of Rice Intensification (SRI)

SRI was evolved in the 1980s in Madagascar. This rice cultivation system increases paddy yields up to 15 t/ha even on infertile soils with less irrigation water and without any external additional inputs (Stoop *et al.*, 2002). In this system transplanting of young seedlings is done in a square pattern with wide spacing, use organic fertilizers and hand weeding, and keeping the field soil moist during vegetative phase. Significant changes occur in plant phenotypic structure, function, grain yield and other yield components under SRI method. SRI increases yields by 50–100% with using only half of the water required under conventional method (Uphoff *et al.*, 2010), and no additional external inputs are needed.

A survey conducted by Moser and Barett (2003) on the farmers of Madagascar to investigate farmer implementation of SRI and the result showed that farmers adapted SRI method and this method fit the soil type and availability of irrigation water and labour. They mentioned that farmers were able to increase grain yields and also reduce irrigation water requirement for rice crop by combing AWD with SRI. In another experiment Thiyagarajan *et al.* (2003) mentioned that there was savings in irrigation water by 56% and 50% using conventional and young seedlings, respectively, without any significant effect on yield under SRI. Krupnik *et al.* (2012) concluded that substantial water saving and increase in water productivity could be obtained with SRI, although significant yield increase should not be expected compared to conventional cultivation practices. Further work should be conducted to investigate the mechanisms underlying these results, and to compare SRI's yield and water productivity performance to other water saving rice management systems.

Table 3.2: Comparative advantage of SRI technology over Conventional system of growing Rice

SRI technology	**Conventional system**
Seed required: 3-4 kg/ac	Seed required: 20-25 kg/ac
Seedlings along with nursery soil in and around the root undisturbed in main field are planted in shallow depth at 10x10 cm^2 spacing.	Seedlings are uprooted from the nursery; roots are washed in the nursery itself and transplanted in the main field at 6x6 cm^2 spacing.
2 leaved seedlings of 8-12 days age in *aman* and of 15-18 days age in summer are planted.	4-5 leaved seedlings of 25-30 days age in *aman* and of 40-45 days age in summer are planted.
Transplanting is done by giving only one seedling per hill.	Transplanting is done by giving 3-4 seedlings per hill.
Population: 16 hills/m^2	Population: 33 hills/m^2
Chemical fertilizer used is less.	Chemical fertilizer used is more.
More organic manure is recommended for use in main field.	Large quantity is not compulsorily recommended for use.
Main field is kept alternately wet and dry till ripening stage.	Standing water level is maintained up to attainment of ripening stage.
Disease pest attack is less.	Pest infestation is more.
Weed infestation is more; chemical and brown manuring is more effective.	Standing water level and puddling do not allow vigorous growth of various weeds.

F. Integrated Crop Management (ICM) - an Alternative to SRI

To overcome the above mentioned limitations farmers can go for Integrated Crop Management (ICM). In this method, 15 - 20 days old rice seedlings are transplanted with wide spacing of 20cm x 20cm or 25cm x 25cm. Here for seedling preparation modified mat nursery is practiced. Combined application of organic manures and chemical fertilizers results higher grain yield. Adoption of integrated pest management (IPM) is also very effective for pest and disease management under ICM as a few incidences of disease and pest attacks is observed due to healthy plants. Thus farmer can reduce or avoid pesticide application.

Advantages of ICM

- Seed requirement for seedling making is less than conventional cultivation practice

- Saving of water as intermittent irrigation is applied
- Reduce cost of external inputs as organic manures are used
- Ideal for high rainfall regions like North-East India
- It is comparatively more easy to handle 15-20 days old seedlings than 10-12 days old seedlings
- A few incidence of pest and disease attacks as the field soil is allowed to dry intermittently
- Higher grain yield due to increase in number of tillering, panicle length and grain weight.

G. Azolla in rice production

Application of urea in rice field is slowly turning out to be toxic with environmental hazards. Beside these, frequent application of urea in crop increase the input cost. In this scenario, eco-friendly Azolla most useful and effective as bio-fertilizer in rice crop to fix the atmospheric nitrogen in association with blue green algae. Utilisation of Azolla as a bio-fertilizer can be taken up as entrepreneurship opportunity. 10-15 tons of fresh Azolla is sufficient for one hectare of rice field, which can provide around 20-40 kg N/ ha/ 20-75 days with increasing grain yield by 10-30%.

Figure 3.5: Use of azolla in rice production

H. Site specific nutrient management (SSNM)

Real time application of fertilizers is a key to improve nutrient use efficiency. Since nitrogen requirement for crops is closely related to crop yield as nitrogen application increase yield. SSSM depends on climatic condition particularly solar radiation, crop management practices, supply of nutrients and dynamics N adjustment of real time. We easily find nitrogen requirement of any crop by using SPAD meter, green seeker (GS) and leaf colour chart (LCC); and nitrogen can be applied according to need of the crop plants. These technologies provide instantaneous results and have been demonstrated as effective tool to schedule N fertilization to crops like rice.

I. Leaf Colour Chart (LCC)

Leaf Colour Chart has been developed in Japan and it is non-destructive method where green colour intensity of rice leaves is measured to identify the nitrogen requirements by crop plants (Nachimuthu *et al*., 2007). This method is also standardized with chlorophyll meter. In both hybrid and inbred rice cultivation, nitrogen application through LCC has been proved superior to locally recommended dose of nitrogen application by many scientists. It was found that when nitrogen fertilizer is applied as per LCC values, farmers can reduce nitrogen application by 20-30 kg/ha without sacrificing any significant rice yield. N application at LCC value less than 3 in Basmati and at LCC value less than 4 in hybrid rice was found beneficial. Moreover, in LCC based nitrogen management, farmers can also skip basal application of N without any disadvantage in terms of grain yield. Sharma *et al.* (2008) mentioned that 10 to 15 % nitrogen fertilizer can be saved by using LCC.

J. Laser land leveller

Laser levelling is a process where the land surface is smoothen using laser equipped drag buckets to maintain proper land levelling. Precision land levelling alters the field to create a constant slope of 0 to 0.2%. Large horsepower tractors or soil movers operated with global positioning systems (GPS) or laser-guided instrumentation are used to create constant slope either by cutting or filling the soilsurface. Various studies suggested that laser land levelling can reduce irrigation water application by almost 25% and increase wheat yield by 30% as compared to conventional practices. Similar increase in yield and reduction in irrigation water application can be seen in case of maize cultivation under zero tillage and laser land levelling. Jat *et al.* (2009) also mentioned that the use of RCTs, including zero tillage and laser land levelling reduced irrigation water application by 23 to 45 % while increasing crop yield.

3.4 Conclusion

Rice, the staple food crop for South Asia is contributing immensely to fill the increasing population but has consequently led to many issues like declining water resources, degrading soil health and environment degradation. Rice cultivation is further responsible for decrease in land and water productivity. Many alternate tillage practices and rice cultivation methods are already invented, tested and recommended for sustainable cultivation of rice so that land and water productivity, soil and environment health must be improved for improving livelihoods of the farmers of South Asia. However, performance of these technologies depends upon soil type and agro- climatic conditions. So, farmers must pick new cultivation methods according to the soil texture and agro-climatic condition. Conventional age-old practices which are responsible for many unsustainability issues must be replaced with advanced and sustainable resource conservation technologies (RCTs). Adoption of these technologies will help to conserve resources sustainably at the field level while to increase crop productivity at the same time. Therefore, these advanced RCTs to gain sustainable food production with minimal harmful impacts on the soil and atmosphere health, underground water resources and improve land and water productivity become more important now than ever to maintain a sustainable livelihoods.

References

Chutiwat, W. and Direk, I. (1997). Effect of rice straw management on soil fertility and rice yield (in Thai). Naresuan J 1(3):30–35.

Kanika, N. (1998) Improvement of paddy soil. In: Technology for Hom Mali rice production (in Thai). Department of Agriculture, Ministry of Agriculture and Cooperatives, pp 33–45.

Mongko,l T. and Anan, P. (2006). Improvement of paddy soil for organic rice (in Thai). Department of Extension Cooperative, Ministry of Agriculture and cooperatives, 22pp.

Prasert, S. and Vitaya S. (1993). Using organic fertilizer to improve paddy soil in Northeast Thailand. In: Proceeding of developing rice and cereal crops in Northeast Thailand (in Thai). Ubonratchatani Rice Research Center, Department of Agriculture, Ministry of Agriculture and Cooperatives, pp 49–71.

Singh, D. D. (2003). Management of crop residue in summer rice and its effect on the soil properties and crop yield. Crop Res 25 (1) : 191-193.

4

Application of Resource Conserving Technology (RCT) in Wheat

4.1 Introduction

Wheat (*Triticum aestivum* L.) is a major food security crop in South Asia, with 42% of the wheat in this region grown following rice (*Oryza sativa* L.) on 13.5 million hectares of land. Resource-conserving technologies offer considerable potential for improving the sustainability and productivity of wheat in the rice-wheat system of South Asia. These include raising input-use efficiency, cutting costs, providing various environmental benefits, and ultimately improving farmer livelihoods and helping to reduce poverty. The conservation technologies described in this paper are in the form of reduced tillage, zero-tillage, and bed planting for wheat in the rice-wheat system. These technologies effectively attempt to solve the problems of late wheat planting and poor plant stands, which limit yields under rice-wheat rotations. Reduced- and zero-tillage practices are discussed in light of problems that warrant additional research, such as interactions between tillage and fertilizer use (methods and timing of application), residue management, development of improved equipment, pest management strategies, variety, and interactions with crop establishment practices for rice. Farmer response to these technologies has been very favorable and area has risen from a few hectares in 1998 to more than 100,000 ha in 2001. Continued acceleration of adoption is expected as local manufacturers keep up with equipment demand. This paper lists the large number of benefits explained by farmers for the new technologies. Resource-conserving technologies are a key to ensuring sustainable food production in South Asia in the next decade.

4.2 Conservation Agriculture (CA)

The term 'Conservation Agriculture' means raising of crops without tilling the soil retaining crop residues on the soil surface. In other words, Conservation Agriculture (CA) refers to the agriculture which is built on conservation tillage by reducing surface tillage even more. The term 'Conservation Agriculture' is coined by the Food and Agriculture Organization (FAO) twenty years back and it has three important principles as follows.

A. **Reduction in tillage:** The prime objective of CAis to reduce tillage operations. This sowing system does not normally disturb 20 to 25% of the soil surface. ZT reduces soil disturbance, energy use and production costs; and increase profit.

B. **Retention crop residues and surface covers on the soil surface:** The another objective of CA is the retention of sufficient amount of crop residues on the soil surface to protect the soil from water and wind erosion; to reduce evaporation loss of soil water, to check run-off and to improve soil productivity by enhancing soil physical, chemical and biological properties for gaining long term sustainable productivity. The amount of residues necessary to achieve these ends will vary depending on the biophysical conditions, the cropping system, and farmer needs and circumstances.

C. **Use of crop rotation:** Adoption of diversified crop rotations helps to minimize weed attacks, insect pest and disease infestation; to utilize the beneficial effects of some crops on the productivity of subsequent crops; and to provide the farmers an economically viable cropping system that minimizes risk (Hobbs, 2010).

4.3 Conservation Agriculture (CA) in India

Intensive efforts for suitable adoption and self-styled promotion of modern concept of conservation agriculture have been undertaken in India for nearly last one and half decades and it is only in the last few years our farmers have started to think of and accept the technologies. An eco-regional initiative taken by the Consultative Group on International Agricultural Research (CGIAR) and the National Agricultural Research Systems of India, Pakistan, Bangladesh and Nepal, namely, Rice-Wheat Consortium for Indo-Gangetic Plains (RWC) has spear-headed our efforts in taking these technologies.

Main concerns

(a) There are long gap between ability and actual productivity of crops.

(b) Productivity has reached to a more or less static status.

(c) Production cost is increasing day by day.

(d) Water table is declining with the march of time due to intensive agriculture, civilization, climatic and other reasons.

(e) Ecological and environmental wear and tears are coming out with greater intensity.

(f) Deterioration of soil health and quality of other natural resources.

4.4 Zero tillage in wheat production

Surface seeding is the simplest zero-tillage system being promoted. In this tillage option, wheat seed is placed onto a saturated soil surface without any land preparation. This is a traditional farmer practice for wheat, legume, and other crop establishment in parts of eastern India and Bangladesh (Catling *et al.*, 1983). Wheat seed is either broadcast before the rice crop is harvested (relay-planted) or after harvest.

Promotion of surface seeding to plant wheat has been done for several years in areas where the soils are fine-textured and drain poorly and where land preparation is difficult and often results in a cloddy tilth. The key to success with this system is having the correct soil moisture at seeding. Too little moisture results in poor germination and too much moisture can cause seed to rot. A saturated soil surface is best. The roots germinate into the moist soil and follow the saturation fringe as it drains down the soil profile. The high soil moisture reduces soil strength and thus eliminates the need for tillage. Once the root germinates and extends into the soil, the root can follow the saturation fringe and still get sufficient oxygen for growth.

As long as soil moisture can be manipulated, surface seeding is also successful on coarse textured soils. There must be enough surface moisture to germinate the seed (soaking the seed can help), and soil strength at the root penetration stage must be low (moisture-dependent) to allow root growth. This may require an early light irrigation on coarse-textured soils. Some farmers who relay wheat into the standing rice crop, place the cut rice bundles on the ground after harvest. This allows the rice to dry but also acts as mulch, keeping the soil surface moist and ensuring good wheat rooting. In China, where surface seeding is also practiced, farmers apply cut straw to mulch the soil, reduce evaporative losses of moisture, and control weeds (Yonglu *et al.*, 2000; Gupta *et al.*, 2000). The standing stubble also protects the young seedlings from birds. However, relay planting can be done only if the soil moisture is correct for planting at this stage.

Figure 4.1: Wheat cultivation with Zero Tillage

4.5 How ZT is advantageous in wheat

- As wheat seeds are sown directly in uncultivated field in lines with Zero Tillage machine ploughing cost is reduced.
- In time seeding (at least 15 days earlier as compared to conventional method) is possible particularly in field after harvest of *aman* paddy.
- Germination of seeds is uniform as proper residual soil moisture is maintained.
- Seed rate is low (100-115 kg/ha) as compared to conventional sowing method (120-150 kg/ha).
- Requirement of irrigation water in ZT is about 20- 40% less than conventional system.
- Fertilizer use efficiency is more as fertilizer is applied by the side of the seed with Zero Tillage machine.
- Weed infestation of *Phalaris minor* and *Chenopodium album* is low. Reduced tillage does not allow the seeds of weeds to generate dense weed population in the field.
- Arsenic contamination of crop: It has been reported by Nora *et al.* (2005) that accumulation of arsenic in wheat on dry weight basis was recorded as high as 0.74 mg/kg in West Bengal. Reduction in use of ground water as irrigation water in this crop under ZT technology is the very good management option to mitigate this problem. Sowing is done in sufficiently moist soil after paddy and thus there is no need of sowing time irrigation as compared to traditional method. Moreover,

wheat field under ZT needs to be irrigated for only 1.5-2.0 hrs for one acre of land owing to conservation tillage practice followed whereas watering for 4.0-5.0 hrs is required in traditional method. Consequently, use of ground water can be reduced by about 20-40% in this technology and risk of arsenic can be lessened to a large extent.

- Sowing of seeds in line is done automatically with this Zero Tillage machine
- Soil properties and soil health are also improved under ZT practice.

Table 4.1: Soil property as influenced by the Zero Tillage over Conventional Tillage

Cultivation technique	pH	OC (%)	P_2O_5 (Kg/Ha)	K_2O (Kg/Ha)	Bulk Density	E.C.	Moisture (%)
CS	7.1	0.4	44.6	311.50	1.4	0.3	9.5
ZT	7.0	0.4	106.2	223.1	1.7	0.1	12.4

Source: Bhattacharyya (2009)

*CS= conventional system; ZT= zero tillage; OC= organic carbon; E.C.= electrical conductivity

4.6 Package of practices for cultivation of Wheat through Zero Tillage

i. **Varieties:** Wheat varieties with vigorous early growth habit and profuse tillering capacity are very good for cultivation under ZT. Varieties like Sonali, HP 1731, HP1737, PBW 343 and PBW 443 are found most suitable over old varieties. If, no special varieties are not available to cultivate wheat under ZT then the varieties that have performed well in conventional method can be used for cultivation.

ii. **Seed treatment:** Certified wheat seed having at least 85% germination capacity should be used for cultivation. Seeds should be mature, free from weed seeds and disease. Before sowing, seeds should be treated with Thiram @ 3.0 g/kg seed or Carbandazime @ 2.5 g/kg seed to protect seeds from seed borne diseases like stem rot or seedling blight. In the areas where incidence of loose smut disease is frequent, seeds seed treatment with Carboxin @ 2.5 g/kg seed should be done before sowing. In the areas where ears become shrinkage due to nematode attack and grains are deformed into wart like structure called 'ear cockle', seeds must be treated with saline water (50 g table salt in one litre of water) followed by discarding the floating seeds and drying of good seeds before sowing.

iii. **Seed rate:** The recommendation seed rate of wheat for ZT is 100-115 kg/ha. More seed rate causes lodging of wheat plants resulting less productivity. Wheat can give higher grain yield with low seed rate due to high tillering capacity, improved size of ear head and more number of grains per head. If the germination percentage of the wheat seeds is less than 85 % then seed rate should be increased. To check the germination percentage of the seeds, place 100 seeds on wet newspaper, roll it up and close the ends. Keep the rolled newspaper under moderate temperature for 3 - 5 days and maintain moisture by spreading water time by time. Open the newspaper roll, counts the germinated seeds and the number is the germination percentage of the seed.

iv. **Soil moisture status:** When wheat is cultivated under ZT, the soil should not be too wet or too dry at the sowing time. High soil moisture content at the sowing time creates aeration stress which causes root rot and poor plant stand. In case of sandy soil one light irrigation after the sowing should be provided in absence of sufficient soil moisture. Sowing of seeds can be done by Zero Tillage machine which is actually a zero-till-seed-cum-fertilizer drill. It is a conventional seed drill fitted with modified furrow openers with sharp edged, a calibrated engraved disc and a cup mechanism for dropping fertilizer. The tractor drawn domestically developed machine opens a number of narrow slits (6-11) for placing the seeds and fertilizer at a depth of 7.5-10.0 cm into the soil. Under the purview of conservation agriculture, zero tillage implies retention of crop residue as mulch on the soil surface and its year round application and management to all crops in the cropping cycle.

Following operations should be practised in order to use zero tillage in right way

(a) **Checking of machinery-** The machine should be checked before use to ensure that all the parts like boxes for placing seeds and fertilizers, openers, chains, nuts etc. are in good condition.

(b) **Calibration-** The machine should be calibrated for both seed and fertilizer properly on the eve of planting season. Sometimes calibration values are mentioned on the body of the machine by the manufacturer as helping tool. To ensure that the machine is supplying the correct quantity of seed and fertilizer, a plastic or paper bag is placed over the spout to collect the dropped materials over certain length or area. The amount of input dropped on unit area is then calculated and compared with the recommended value given in label on the machine. Adjustments are done if required and the drill is calibrated again until satisfactory performance is achieved. The germination percentage of the seed should be considered and rate of seed should be adjusted accordingly.

(c) Adjustment of the drill- The three point hitch by which the drill is fixed to the tractor should be adjusted. The drill should be levelled from side to side and have just ample forward and backward adjustment to enter the soil at proper angle. It should not be adjusted too steep or too shallow as planting will be too deep or dropped on the surface, respectively. One man can remain sited at the time of seeding to keep constant watch whether the holes for dropping seeds and fertilizers are clear and make it clear as and when required. Keeping the speed at 4-5 km/hr, one tractor drawn seed drill can sow one acre of wheat plot in an hour.

Figure 4.2: Zero-till planting of wheat

v. Application of fertilizer

In ZT wheat, N, P and K nutrients are applied @ 120:60:60 kg/ha respectively. Time of application to be followed is: (a) ½ N + whole P and K as basal, (b) ¼ N as 1st Top dressing (21 DAS) and (c) ¼ N as 2nd Top dressing (42-45 DAS). Once the drill calibration is over, diammonium phosphate at required quantity for P and N or 10-26-26 fertilizer for N, P and K at basal dose should be placed in the fertilizer box and drilled into the soil at 5 cm depth by side of the seed. Remaining nitrogen should be applied through urea.

Above fertilizers should not be mixed and the fertilizer should not be left in the box for long time because it forms cake and free flow of fertilizers is disturbed while dropping the same into the soil. Urea should not be applied during seedling stage as it burns roots of young seedlings. In case of light sandy soils, fertilizer application should be done after irrigating the plots to get better fertilizer use efficiency.

If the symptoms of zinc deficiency is seen in the wheat field application of Zink-EDTA @ 0.5 g/litre of water or zinc sulphate @ 5g + lime @ 2.5 g/litre of water at 25- 30 DAS and 45-50 DAS should be done. In the areas where boron deficiency is found Borax (Di-sodium octa-borate tetra-hydrate) @ 1.5-2.0 g/litre of water should be applied to reduce shortage of micronutrients and improve yield.

vi. Water management

Generally, four irrigations at crown root initiation stage (21 DAS), maximum tillering stage (42-45 DAS), early flowering stage (75-80 DAS) and grain filling stage (100-110 DAS) should be provided. Irrigation scheduling mostly depends on moisture status and texture of the soil. In case of light textured soil irrigation should be given earlier to facilitate good root growth and subsequent tillering. However, depth of irrigation in ZT is comparatively less than that in conventional tillage method, because soil surface is not disturbed in whole field and tillage is reduced to the maximum limit. Time required to irrigate the plot is also reduced to the large extent as undisturbed soil surface allows the water to flow over and reach all the places in the plot easily in less time than conventional system. Only 1.5 - 2 hours are enough for irrigating one acre (0.4 ha) wheat field under ZT, whereas 4 - 5 hours is needed to irrigate in conventional system.

vii. Weed management

Major weeds found in wheat field are as follows.

Broad leaved: Bathua (*Chenopodium album*), Wild spinach (*Gnaphalium purpureum*, *Gnaphalium indicum*), Gajri (*Fumaria parviflora*), Krishnaneel (*Anagalis arvensis*), clover (*Melilotus indica*) etc.

Grassy: Wild oat (*Avena fatua*), Phalagrass (*Phalaris minor*), Shyama grass (*Echinocloa colonum*), Karat grass (*Lolium temulentum*) etc.

Pre-plant (7-10 days before sowing) application of Glyphosate @ 8-10 ml/lit of water (450-550 lit/ha) in morning time is found very effective to control perennial weeds. Pre-emergence spray of Pendimethalin 30 EC @ 4-5 ml/lit of water (600-800 lit/ha) 2-3 days after sowing of seeds is also very effective to control grassy weeds.

Another weedicide 2, 4-D 80% a.i. @ 625 g/ha or 1.0-1.5 g/lit of water (400-600 lit/ha) is also effective within 30-35 DAS. Very early application of this chemical causes malformation of ear, while application beyond 35 DAS is found very less effective for weed control.

Table 4.2: Performance of ZT in wheat cultivation in West Bengal (2005-08)

Sl. No.	District	Average yield (q/ha) ZT method		% increased over conventional
		ZT method	conventional method	
1.	Birbhum	27.2	21.9	24.4
2.	Burdwan	25.2	21.9	15.2
3.	Coochbehar	24.6	22.7	8.8
4.	Dinajpur (South)	23.5	18.9	24.7
5.	Murshidabad	24.6	20.3	21.4
6.	Nadia	26.8	23.6	13.6
7.	Jalpaiguri	34.2	30.1	13.6
	Mean	26.6	22.7	17.0

Source: Bhattacharyya (2009)

viii. Plant Protection

Plant protection measure should be taken as per the need. Major damage of wheat crop is caused by termite. To control termites soil application of Chlorpyriphos 1.5% dust or Folidol 5% dust @ 30 kg/ha should be done during final land preparation. Placing 2.0 gm zinc phosphide with 100 gm of edible flour and edible oil in the field in the evening should be done to control rats. Non-poisonous pre-control baiting should be compulsorily done for few days in order to attract the mammal pest towards the bait containing poison. For getting total control, chemicals like Warfarin, Fumerin, Rodafarin, Ratafin etc. may be used in baits.

Among different diseases rust, smut and leaf blight are important in case of wheat. Brown rust is usually found in the month of February and it delays the emergence of ear heads. Two to three foliar spray of carbendazim 12% + mancozeb 63% @ 2.5 gm/lit water is found effective to protect the crop from rusts. Spraying of 2.0 % urea with fungicide is also very effective for rust control. If black powdery spores smut disease is found in the ear head, the infected ear head should be wrapped in moist cloth, detached from the plant and burnt to avoid further spread. Seed treatment must be done to control seed borne diseases like smut and leaf blight in wheat.

ix. Yield

Average 10 to 20% increase in crop productivity can be achieved through zero tillage cultivation of wheat over conventional method of cultivation. Yield advantage can be even more if the ZT practice goes in favour of wheat cultivation.

4.7 Case study

The study was conducted in Kaithal district of Haryana during 2014-15 in nine villages with 120 purposively selected farmers who have adopted either of the three resource conservation technologies viz;zero tillage, rotary tillage and laser land leveller or in combination. Majority (68.98 %) of the farmers belonged to middle age group (31-50 years) followed by young (20.33 %) and old (18.69 %). Majority (85 %) of the farmers were literate and agriculture was their main occupation. Farmers were categorised in four groups on land holding basis and 6.5 percent, 15.46 percent, 39.02 percent and 39.02 percent, in marginal, small, medium and large categories, respectively. In the study area, majority of the farmers preferred Agriculture Development Officers (ADOs) as main source of information. The study on adoption pattern of resource conservation technologies revealed that 69.92 percent of the farmers had adopted zero tillage, 51.22 percent rotary tillage, 67.48 percent laser land leveller, 22.76 percent zero tillage+rotary tillage, 47.15 percent zero tillage+laser land leveller, 33.33 percent rotary tillage+laser land leveller and 13.01 percent zero tillage+ rotary tillage+ laser land leveller. Majority (84.88 %) of the farmers used zero tillage for timely sowing and 40.70 percent used for both i.e. timely and late sown conditions. Regarding the impact of zero tillage technology, majority of the farmers (77.91 %) told that there was saving of time under zero tillage sowing as compared to conventional tillage. Majority of the farmers (80.23 %) told that there was an increase in broader leaf weeds under zero tillage and reduction in narrow leaf weed population (80.23 %) and 76.74 percent of them recorded reduction in overall weed population. All the farmers agreed that there was cost saving under zero tillage technology. Most of them (91.86 %) harvested more yield as compared to conventional tillage. Alarge number of farmers (61.63 %) observed longer crop duration (4-7 days), increased soil fertility (51.16 %), increased organic carbon in the soil (54.65 %) and increased moisture retention capacity of soil (50 %) and less lodging (87.21 %) as compared to conventional tillage. All the farmers agreed that there was time saving under rotary tillage in wheat sowing. Majority of them (68.25 %) recorded decrease in cost of cultivation, more yield (41.27 %), increased soil fertility (86.96 %), increase in organic carbon content of soil (57.14 %), increase in moisture retention capacity of soil (52.33 %), avoidance of terminal heat (85.71 %) and increase in lodging (46.03 %). Due to adoption of laser land leveller farmers (45.78 %) could save time in field preparation, 2-3 percent increase in cultivable area, less time required to irrigate the crop, 30-40 percent water saving, more yield (96.39 %) and less lodging (95.18 %).

4.8 Conclusion

Resource conservation technologies in wheat improve production at a lower cost by improving the efficiency of natural resources, protecting the environment and improving livelihoods of farmers, and ultimately help in alleviating poverty. Though, the success of these technologies depends on rapid adoption by the farmers. Accelerated adoption was mainly the result of the change in paradigm for extending the technology. Instead of the linear approach to extension commonly found in the Green Revolution era, the importance of partnerships, expanded stakeholders, and participatory approaches in which farmers could experiment and feed back information soon became apparent. RCTs are the key to ensure sustainable food production in South Asia in the next decade to meet food security. Overcoming mind-sets that hold traditional beliefs about excessive tillage and providing the enabling factors that allow exposure of the technology to all those involved in agriculture will be key factors for future success. This technology revolution is seen as one way to sustainably increase food production to meet future demand while conserving natural resources, improving farmer livelihoods, and reducing negative effects on the environment. Of course, all of these benefits will be of little use unless the nations in South Asia control their population growth.

References

Abrol IP and Sangar Sunita. (2006). Sustaining Indian agriculture, conservation agriculture the way forward. Current Science. 91(8): 1020- 1025.

Bhattacharyya P, Mallick EH, Pathak H and Ladha JK. (2008). Wheat in West Bengal. Directorate of Agriculture, Govt. of West Bengal and Rice-Wheat Consortium for Indo-Ganjetic Plains. New Delhi. P 35.

Bhattacharyya P. (2009). Impact of zero tillage technology on yield of wheat and direct seeded rice and soil properties. Satsa Mukhapatra- Annual Technical Issue. 13: 6-12.

Blanco- Canqui H and Lal R. (2008). No tillage and soil-profile carbon sequestration: an on farm assessment. Soil Science Society of American Journal. 72: 693-701.

Hobbs PR. (2010). Increasing the efficiency of food production: the need to adopt resource conserving technologies in India. Satsa Mukhapatra- Annual Technical Issue. 14: 01-17.

Nora S, Berner ZA, Agarwala P, Wagner F, Chandrasekharam D and Stuben D. (2005). Impact of irrigation with As rich ground water on soil and crops: a geochemical case study in West Bengal delta plain, India. Applied Geochemistry. 20:1890- 1906.

Sharma SN and Prasad R. (1999). Effect of summer legume on ricewheat cropping system. Bioresource Technology 67: 171-175.

Tandon, HLS. (2004). Fertilizer in Indian agriculture- from 20th to 21st century. FDCO, New Delhi, India. p 240.

5

Resource Conservation Technology in Pulse Based Cropping System

5.1 Introduction

Indian agriculture is now at crossroads. We have come a long way. Over the past five decades our strategies, policies and actions were guided by goals of 'self sufficiency' in foodgrain production with main focus on rice and wheat. Indian agriculture has been successful in achieving higher production of food grains to secure food security of the nation. While, the success for increasing food grain production has been achieved without major jump in the production and productivity of pulses in the country, this was accompanied by widespread problems of resource degradation, which now poses serious challenges to meet the demands of an increasing population and lifting our people above the poverty line.

Rice and wheat are two major food grains of India. Therefore, primary focus is given on improving and sustaining their productivity. With the advent of the "Green Revolution", these two crops have come to occupy a significant area in the country. Rice and wheat have similar adventitious rooting systems and they are heavy feeders of plant nutrients, and rice requires a large number of irrigations (20 to 25 irrigations in the irrigated rice-wheat areas on light alluvial soils). Rice and wheat exploit the soil to a greater extent making it poorer year after year. The rice-wheat based cropping system annually removes more than 650 kg of N, P and K from one hectare soil resulting negative balance of these nutrients even with use of recommended fertilizers which adversely affect the sustainability of production system (Abrol *et al.*, 1999).

Pulses are next to cereals in terms of their economic values and nutritional importance as human food. The ability of pulses to fix atmospheric nitrogen in the soil-crop system is their unique and beneficial characteristics among all plant species. Thus, pulse crops can contribute to achieve the objectives of increasing productivity and improving the sustainability of rice and wheat based cropping systems. The pulses grown together with or before a cereal crop can reduce and sometimes substitute the need of nitrogen application as well. The deleterious effect of spring fallow on soil physical and chemical properties, soil organic carbon content and soil fertility can be decreased by cultivating pulse crops during fallow period.

In spite of many beneficial effects, pulses are still cultivated as secondary crops with respect to management practices and have always been relegated to marginal lands and cultivated with minimum or no inputs. Pulses are grown on marginal and degraded lands over the years with minimum or no inputs. Resource conservation practices have shown advantages over traditional cultivation practices by means of improving the productivity and soil health.

5.2 Conservation Agriculture vs. Resource Conservation Technology

According to FAO, Conservation Agriculture (CA) is an approach to manage agro-ecosystem for gaining sustained productivity, higher profits and food security of nation while preserving the resource base and the environment. CA is characterized by three principles:

i. Continuous minimum mechanical soil disturbance,
ii. Permanent organic soil cover, and
iii. Diversification of crop species grown in sequence or association.

However, the term resource conservation technology (RCT) refers to the practice that enhances resource or input use efficiency by introducing zero or reduced tillage, laser land levelling, bed and furrow configuration for planting crops etc.

5.3 Advantages of conservation agriculture

- Reduced soil erosion by wind and water
- Increased water infiltration and storage
- Reduced evaporation
- Prevents overheating of the soil surface which favours seed germination
- Build-up of SOM
- Improved aggregate stability and soil structure
- Deepening of rooting horizon
- Less drought susceptibility
- Better quality of rural life
- Improvement in soil quality and fertilizer efficiency

- Reduction in poverty
- Reduction in labour demand for small framers.

5.4 Pulses: Component of RCT

A. Sequential Cropping

Sequential cropping is a kind of multiple cropping system in which two or more crops are grown in a sequence on same land within a crop year, one crop being sown after the harvest of the other. Pulse crops are specially introduced in the cereal or oilseed based cropping systems to break the monotony of the crop rotation. Pulses are mainly rain-fed crops, require fewer inputs and provide safeguard against crop failure due to shortage of rainfall and other weather hazards. Introduction of pulse crops in the cropping system also improves nitrogen content of the soil.

Besides improving the soil fertility, inclusion of pulse crops in rice-wheat cropping system increases the system productivity. Inclusion of *rabi* pulses significantly increased the productivity of succeeding rice crop in comparison to wheat. In low land areas of eastern UP, Bihar and parts of MP, where excess moisture after harvest of rice poses serious problem, rust resistant high yielding lentil varieties have been found suitable for rice-lentil sequence. However, in relay cropping system rice - lathyrus have been found most remunerative, followed by rice-linseed. Short duration high yielding varieties like Amber, Uday, HUR 15 and HUR 137 of rajma have been successfully introduced in the central and eastern plain of India (Table 5.1)

Table 5.1: Nitrogen economy due to inclusion of pulses in sequential cropping

Preceding pulse crop	Following cereal	Fertilizer N- equivalent (kg N/ha)
Chickpea	Maize	60
Chickpea	Rice	40
Pigeon pea	Wheat	40
Mungbean	Rice	40
Urdbean/mungbean	Wheat	30
Lentil	Maize	30
Fieldpea	maize	25
Rajmash	Rice	10
Cowpea	Rice	40
Cowpea	Wheat	43

Source: Subbarao, 1988

B. Intercropping

Intercropping is the practice where two or more crops are grown in proximity. The most common goal of intercropping is to produce a greater yield on a given piece of land by utilizing resources that would not be utilized by a single crop. In India under rain-fed condition pulse crops are generally intercropped with coarse cereals or oilseed crops. The crops with slow initial growth habits and deep tap root systems are more suitable for intercropping with cereal crops and oilseed crops than any other crops. Apart from increasing the soil fertility, higher resource use efficiency and yield profitability, pulses as intercrop also pay a vital role as safeguard under moisture stress condition. Studies conducted at IIPR have also revealed that introduction of pulses as intercrop could reduce the disease and weed intensity. The major intercropping systems involving pulses are as follows:

i. Kharif pulses

The early maturing cereal crops like maize, sorghum and millets offer a good option for intercropping with long duration pulse crops like pigeonpea as they utilize resources and accumulate dry matter during the initial growth period of pigeon pea. The reproductive phase of the intercrops does not coincide with that of pigeonpea and as a result the yields of cereal crops are not affected. After harvesting of cereal crops, the growth of pigeonpea is compensated and additional yield of pigeonpea is obtained. Pigeonpea + sorghum intercropping system with spatial arrangement of 2:1 row ratio recorded higher pigeonpea equivalent yield and B: C ratio than 1:1 and mixed planting system. In early pigeonpea + groundnut intercropping system, 2:2 row ratio with North–South row orientation was found most productive (1.24 kg pigeonpea equivalent yield/ha) and increased land use efficiency by 90%. The companion crops viz., mungbean, urdbean, soybean, cowpea and sorghum of pigeonpea appreciably suppressed the weed flora under both uniform and paired row planting conditions. However, weed suppression under uniform row planting (1:1) was higher than paired row planting due to close sowing. Among different companion crops, cowpea was the most effective under uniform row planting. Pearl millet + mungbean and pearl millet + urdbean particularly under paired row system (2:1) are also highly productive and efficient.

ii. Rabi pulses

Lentil, chickpea, and rajma also provide suitable options for intercropping with cereals and oilseeds. Kabuli chickpea + barley in 3:1 row ratio recorded maximum productivity and higher LER (1.3) as compared to sole crops. Similarly, chickpea + linseed intercropping was found highly productive and profitable than their sole cropping.

iii. Spring season

Spring sown sunflower + mungbean in 6:2 row ratio was found most efficient intercropping system. Intercropping of mungbean variety PS 16 was found most compatible with sunflowers cv. Modern for intercropping. Similarly, due to slow growth rate and wider row spacing, spring planted sugarcane provides a good opportunity to grow short duration low statured plants like mungbean and urdbean. Mungbean has advantage over urdbean and mungbean varieties for intercropping with spring sugarcane.

C. Improving soil health

In rice-wheat cropping system, major concern for sustainability is decline of soil organic carbon (SOC). The process of decline in SOC can be reversed by inclusion of pulses in the cereal based system. The improvement in SOC was found under pigeonpea- wheat, rice-lentil and rice-wheat-mungbean cropping systems. Similar observation was also observed in a long-term field experiment at IIPR in which SOC improvement was found in rice-chickpea and rice-wheat-mungbean cropping systems in lowland situation and maize-chickpea, pigeonpea-wheat and maize-wheat-mungbean cropping systems in comparison to rice-wheat and maize-wheat copping system respectively.

The inclusion of pulses in the cropping system not only increases the nutrient content of the soil but also brings positive changes in physical and biological properties of the soil, resulting in buildup of soil fertility. In a long-term maize based field experiment, improvement in available P, K, S, Zn and B content in the soil was recorded due to inclusion of pulses in the cropping system (Table 5.2). Similar trend was also observed in rice based cropping system in low land.

Table 5.2: Effects of pulse based cropping systems on soil chemical properties in long-term experiment

Cropping system	Available P (kg/ha)	Available K (kg/ha)	Available S (kg/ha)	DTPA –Zn (kg/ha)	B (kg/ha)
Maize - wheat	16.00	173.00	17.30	0.60	0.90
Maize – wheat - mungbean	17.20	186.00	19.40	1.1	0.90
Maize - wheat – maize - chickpea	18.00	185.90	18.50	0.80	1.00
Pigeonpea - wheat	16.80	183.20	19.10	0.80	1.00
Rice - wheat	18.55	234.20	14.10	1.68	0.86
Rice – wheat - mungbean	18.37	271.60	16.71	1.60	0.89
Rice – wheat - chickpea	21.20	247.90	17.54	1.69	0.92
Rice - chickpea	21.55	243.40	17.15	1.82	0.93

Source: IIPR Annual Report, 2011-12

Pulses are known to improve microbial activities in the soil. Pulses release unused fixed nitrogen to soil which adds to soil by microbial activities. The increase in nitrogen also helps in improving microbial activities to breakdown carbon rich residue of non-leguminous crop. Improvement in soil microbial biomass carbon is also recorded in a long-term study (Table 5.3). Dehydrogenase activity, an index of soil microbial activity was seen higher after the harvest of legumes in comparison to cereal crops. Thus, pulse crops increase the diversity in soil flora and fauna which lead to greater stability of the total life of the soil.

Table 5.3: Microbial biomass carbon in maize and rice based cropping systems

Cropping system	**Microbial biomass carbon (µg/g)**			
	Control	**Crop residue + biofertilizers + FYM @ 5 t/ha**	**Inorganic fertilizers (N,P,K,S,Zn,B)**	**Mean**
Maize-wheat	247	298	291	279
Maize-wheat-mungbean	327	350	338	338
Maize-wheat-maizechickpea	310	338	334	327
Pigeonpea-wheat	295	305	301	300
Rice-wheat	262	305	300	289
Rice-wheat – mungbean	367	376	361	368
Rice-chickpea-rice – wheat	305	342	358	335
Rice-chickpea	301	336	338	325

Source: Ali and Venkatesh, 2009

5.5 Conservation tillage and pulses

Excessive tillage operation of field may result in short term increase in fertility, but degrade soils in the medium and long run. Structural degradation and erosion of soil, loss of organic matter and falling biodiversity are all to be expected. Soil erosion due to excessive tillage operations has forced us to look for alternatives and to reverse the soil degradation process. This led to promote conservation tillage, and especially zero-tillage in many parts of the world. Over the last two decades the technologies have been improved and accepted by the farmers in our country, mostly in Indo-Gangetic plains. Conservation tillage refers to the planting, growing and harvesting of crops with minimal disturbance to soil surface. Conservation tillage (CT) is designed to reduce erosion and maintain or improve soil health and infiltration by reducing surface sealing and enhancing macro-pores connectivity and flow.

5.6 Residue management and pulses

Crop residues are the parts of plants remaining after harvest. Crop residues are important in nutrient distribution and plant growth as they affect the amount of available soil. Crop residues influence nitrogen cycling in soil because they are the primary sources of C and N in the soil. Residues allow N to be available to plants for longer period of time through initially immobilizing, and then gradually mineralizing N. However, in many parts of the India, crop residues are burnt in the field due to lack of knowledge of farmers about the beneficial values of crop residues and lack of proper technology for in situ incorporation of crop residues (Samra *et al.*, 2003). Burning of one ton rice straw produces 3 kg particulate matter, 60 kg CO, 1460 kg CO_2, 199 kg ash and 2 kg SO_4^{2-}. The heat generated from burning cereal straw can penetrate into the soil up to 1 cm depth and increases the soil temperature to as high as 42.2°C. About 32-76 % of straw weight and 27-73 % N are lost in burning. Bacterial and fungal populations are decreased immediately upon burning.

Pulse crops also contribute to soil organic matter by leaf litter. It is observed that in northern India long duration pigeon pea can add up to 2.8 t/ha leaf litter during whole crop season, whereas chickpea can add up to 1.7 t/ha leaf litter. On an average, nutrients contribution by these leaf litters varies between 8 to15 kg N, 2.5 to 5.0 kg P and 8 to 24 kg K per hectare. Besides adding the fertilizers, incorporation of crop residues also improves soil properties resulting increase in productivity and fertilizer use efficiency. In rice-chickpea sequence, yield of chickpea is significantly influenced by incorporation of rice residue and highest seed yield can be obtained with incorporation of chopped straw + irrigation + 20 kg N/ha. Incorporation of chopped residue of mungbean + irrigation + 20 kg N/ha can give maximum wheat yield.

Incorporation of mungbean and urdbean residues on the field raised the organic carbon content by 35.48% over control. Residue incorporation also increases the availability of N by 24.6 %, P by 11.5 % and K by 18.5 % over the initial nutrient levels. Soil physical parameters such as bulk density, particle density, pore space and water holding capacity also improved under residue incorporation plots (Table 5.4).

Table 5.4: Effect of crop residue incorporation on soil physical properties

Treatments (Residue management)	Bulk Density (g/cc)	Particle Density (g/cc)	Pore space (%)	WHC (%)
Mungbean 1	1.38	2.42	45.5	37.3
Urdbean 1	1.39	2.39	44.65	38.3
Mungbean 2	1.38	2.38	46.80	38.3
Urdbean 2	1.38	2.40	47.00	41.60
Mungbean 3	1.34	2.38	47.32	42.50
Urdbean 3	1.35	2.39	48.23	45.10
Mungbean 4	1.32	2.36	49.63	46.40

Treatments (Residue management)	Bulk Density (g/cc)	Particle Density (g/cc)	Pore space (%)	WHC (%)
Urdbean 4	1.33	2.35	48.20	45.90
Control	1.44	2.50	38.15	33.40
CD (p=0.05)	0.05	0.10	3.51	3.8

Note: 1- Incorporation; 2- Incorporation + irrigation; 3- Chopping + incorporation; 4- Chopping + incorporation + irrigation (*Source*: Singh *et al*., 2012)

Table 5.5: Effect of residue incorporation on periodic changes in microbial biomass carbon and organic carbon content in soil

Treatment	Periodic changes in soil microbial biomass carbon after residue incorporation (µg /100 g)				After wheat harvest	
	7	14	33	56	Microbial biomass carbon (µg /100 g)	Ratio of microbial carbon to soil organic carbon (%)
Mungbean[1]	335	347	345	351	262	6.71
Urdbean[1]	270	178	282	264	222	5.28
Mungbean[2]	345	351	358	367	322	8.25
Urdbean[2]	230	237	237	230	312	7.60
Mungbean[3]	348	369	363	378	327	9.08
Urdbean[3]	330	351	343	351	337	9.10
Mungbean[4]	355	395	377	391	320	9.14
Urdbean[4]	320	359	375	327	347	9.37
Control	240	242	268	248	132	4.25

Note: 1- Incorporation; 2- Incorporation + irrigation; 3- Chopping + incorporation; 4- Chopping + incorporation + irrigation (*Source*: Singh *et al*., 2012)

Technology of raising row crops on beds and furrows system is gaining popularity amongst the progressive farmers, mainly because the cost of crop production is considerably reduced as a result of minimum tillage, water saving, etc. Bed shapers/makers are used behind the tractors to form beds and furrows to sow row crops. Some of the advantages associated with furrow irrigated raised bed (FIRB) technology of crop production are:

- Saving of 30-40% irrigation water.
- Reduce seed requirement and fertilizers (by 25-30%) as compared with flat system.
- Reduce chances of plant submergence due to excessive rain or over irrigation.
- Less crusting of soil around plants and, therefore, more suitable for saline and sodic soils.

- Adaptable to various crops without changing basic design or farm layout.
- Enhance resource use efficiency due to local application and effective utilization by plants.
- Minimize the chances of lodging of crops, especially during rainy season.

Reduce tillage under permanent beds reduces cost of labour and other inputs like fertilizers and seeds. Provide opportunities to take additional green leafy vegetables in furrows.

A trail conducted at IIPR reported that in rice-lentil cropping system, raised bed enhanced the yield of lentil by 16.5% over flatbed sowing. Increase in phosphorus from 25 kg/ha to 50 kg/ha did not show significant increase in lentil yield under raised bed as in case of flatbed system of planting. Similarly, lentil yield under 20 kg S/ha was at par with 40 kg S/ha in raised bed and significantly higher in flat bed. The nodulation in lentil was also higher under raised bed. In another set of study raised bed size and number of lines per bed was evaluated. The study revealed that performance of chickpea and urdbean under 75 cm raised bed was better than 67.5 cm flat bed and the 2 lines per bed was at par with 3 lines per bed.

5.7 Farm mechanization and pulses

Farm mechanization is needed to improve the land and labour productivity. The rising wage rate and non-availability of manpower during peak hours further strengthen the case for mechanization. The reduction in time of farm operations and increasing precision in input application with minimum cost demand for more use of farm machinery. Various farm operations such as land preparation, levelling, sowing or planting, irrigation, fertilizer application, plant protection, harvesting and threshing need a high degree of precision to increase the input use efficiency and reduce the losses. For example, sowing of the seeds at proper soil depth and uniform application of fertilizers at required doses can only be possible with proper mechanical devices and when these farm operations are performed through indigenous methods, their efficiency is reduced. The time taken to perform different farm operations is a factor which helps to determine the cropping intensity. To ensure timeliness of various operations, mechanical equipment which have higher output capacity and cut down the number of operations to be performed are necessary. This has helped in increasing area under cultivation and cropping intensity. The benefits of mechanization can be utilized not only by the farmers having large holdings, but even farmers with small holdings can utilize selected improved farm equipment through custom hiring to increase productivity and reduce cost of production.

The development, standardization and adoption of suitable farm machinery for conservation agriculture including seeding with the minimum soil disturbance keeping residue maintained on the soil surface and developing harvesting equipment are of major concern. To fetch the benefits of CA, the machinery equipment suitable to CA must be adapted to small farmers under different ecological conditions. The main criteria which should be considered during the development of this machinery equipment are that the machines should be suitable for the use in small farms, inexpensive, easily maintainable and repairable and environmental friendly. The role of farm mechanization in CA especially in rainfed regions would need to be very specific and could include:

- Suitable equipment which can work on unploughed soil.
- Direct seeder need to be designed to seed in the surface mulch in untilled soils.
- Specific equipment for sowing of specific crop under CA need to be available for farmers.

A. Raised bed planter

The ridge or raised bed planting of crops not only ensures desired plant stand of the crop but also minimizes the incidence of phytophthora stem blight due to the proper drainage system. It removes the negative effects of water logging. This method increases water use efficiency by saving irrigation water because in this method water is applied only in the furrows requiring less volume of water and saves 20-30% water compared to flat method. The other advantages of raise bed technology include:

- Save seed and fertilizer by 25%
- Less problem of *Phalaris minor*
- Mechanical weed control between the rows
- Less lodging of crops
- Promotion of paracropping i.e., moong/green manure crops in furrow before wheat harvest
- Diversified application for legume crops etc.

Figure 5.1: Raised bed planter

B. Happy seeder

Happy seeder was developed by PAU, Ludhiana, in collaboration with Australian Centre for International Agricultural Research was developed for residue covered fields. It is a combination of two machines, a forage harvester and a zero tillage drill. The chopped materials are blown directly behind the drill and float down on the field surface as mulch. The 'Turbo Happy Seeder' is the advanced, modified and light weight version of the 'Happy Seeder' used in presence of loose and or anchored residues.

Figure 5.2: Happy Seeder

C. Star wheels drill

Direct drilling in combine harvested rice field requires drill capable of cutting through large quantities of crop residue, penetrating untilled soil and dropping seed 25 to 50 mm deep. Disc opener worked under trashy condition fails to cut through the crop residues resulting seed being placed either in the crop residue or on the soil surface. In fixed type furrow opener drills, crop residues tend to collect in front of tynes and block the machine operation. Therefore, there is need to develop a device which may cut residue and help the openers to penetrate up to desired depth for seeding. The star wheels drills are used to plant the crop into high density of crop residues in irrigated condition. The main limitation of star wheels drill is improper planting depths. However, this limitation can be managed by using proper irrigation at the time of sowing.

D. Laser land leveler

Land levelling is necessary for good soil and crop management practices. It saves irrigation water, facilitates agronomic practices and increases yield and quality of the produce. Levelled land also helps in smooth operations of various farm machineries during field operations. The main components of laser guided land leveller are Laser transmitter, Laser receiver mounted on electric manual mast and one receiver is used for mapping, control box, electric/ manual mast fitted on the leveller, external hydraulic system of leveller and drag scraper. The advantages of laser land leveller are as follows:

- Precise levelling and smooth surface
- Reduction in time and amount of water required to irrigate the field
- Uniform distribution of water
- Increase crop yields by 10-20%
- Control water logging and soil salinity
- Increase fertilizer uptake efficiency
- Increase machinery use efficiency up to 30%
- Increase cultivable land area by 2-5%.

Figure 5.3: Laser land leveller

E. Direct drilling (no-till drilling)

It is a practical tool to use crop residues and explore its advantages. Rabi pulses after rice are generally delayed due to late harvesting of rice, soil wetness, poor rice residue management and multiplicity of tillage operations, scarcity of power source and appropriate farm equipment during peak hours and sometime even arrivals of rain in the month of November. The major advantages of Zero-till drill include:

- Saving in cost of seed bed preparation by Rs. 2500 to 3000 per ha
- Saving of 37.3 to 70 litters diesel/ha depending on soil condition
- Time saving in seed bed preparation and sowing
- Advancement in sowing time over conventional system by 2 weeks depending upon soil condition
- Saving in water up to 30%

Figure 5.4: Zero Till-cum-Fertilizer Drill

F. Rice fallow and no-till drill

The timely sowing of pulses is crucial in view of the moisture deficit during critical periods in the rice fallow conditions. The farmers usually broadcast the seeds to take advantage of residual moisture in rice fallow. The mechanization in such conditions needs development, standardization and adoption of farm machinery for direct seeding under residue retained on the soil surface. The small landholder farmers, apart from broadcasting, also sow the seeds in the furrow opened with the help of hand plough. The hand plough generally opens trapezoidal shaped wide and shallow furrows in which poor germination and poor plant stand establishment are observed. This practice is also inconvenient, labour intensive and not suitable for rice fallow.

References

Abrol, I.P., 1999. Sustaining Rice-Wheat system productivity in Indo-Gangatic plains: water management related issues. Agricultural Water Management 40, 31-35.

Ali, M. and Venkatesh, M.S. 2009. Pulses in improving soil health. Indian farming 58(11): 18-22.

Annual Reports. 1986-2012. Indian Institute of Pulses Research, Kanpur.

Samra, J.S., Bijay-Singh, Kumar, K., 2003. Managing crop residues in the rice-wheat system of the Indo-Gangetic plain. In: Improving the Productivity and Sustainability of Rice-Wheat Systems: Issues and Impact. ASA, Spec. Publ. 65 (J. K. Ladha, J. E. Hill, J. M. Duxbury, R. K. Gupta, and R. J. Buresh, Eds.).pp 173–195. American Society of Agronomy, Madison, WI.

Singh, K.K. Srinivasarao, C.H., Swarnalakshmi, K., Ganeshamurthy, A.N. and Narendra Kumar. 2012. Influence of legume residues management and nitrogen doses on succeeding wheat yield and soil properties in Indo-Gangetic Plains. Journal of Food Legumes 25(2): 116-120.

Subbarao NS. 1988. Biological nitrogen fixation: Recent developments. Oxford and IBH publication. pp. 1-19.

Twenty five Years of Pulses Research at IIPR. 2009. Indian Institute of Pulses Research, Kanpur.

6

Weed Management in Conservation Agriculture Systems

6.1 Introduction

The birth of modern conventional agriculture coincides with industrial revolution. The identification of N, P and K as critical factors in plant growth led to the manufacture of synthetic fertilizers. While chemical fertilizers and pesticides have existed since the 19th century, their use grew significantly in the early 20 century with the invention of Haber-Bosch method for synthesizing ammonia. The rapid farm mechanization, especially in the form of tractor and combine harvester coupled with science-driven innovative management methods and resources led to the efficiency of enabling outputs of high quality produce per unit area and time. The contribution of Norman Borlaug and other scientists since 1940s towards improvement of crops for increasing crop yields further accelerated the modern agriculture and initiated the era of `Green Revolution'. However, the growth of conventional agriculture was on the basis of capital depletion and massive addition of external inputs such as energy, water, chemicals etc. The transformation of 'traditional animal based subsistence farming' to 'intensive chemical and tractor-based conventional agriculture', has led to multiplicity of issues associated with sustainability of these production practices. Conventional crop production technologies are characterized by: (1) intensive tillage to prepare fine seed- and root-bed for sowing to ensure proper germination and initial vigour, faster absorption of moisture, control of weeds and other pests, mixing of fertilizers and organic manures; (2) mono-cropping systems; (3) clean cultivation involving removal or burning of all residues after harvesting leading to continuous mining of nutrients and moisture from the soil profile; and bare soil with no soil cover; (4) indiscriminate use of insecticides and pesticides, and excessive and imbalanced use of chemical fertilizers leading to decline in input-use efficiency and factor productivity, and increase in pollution of environment, ground water, streams, rivers and oceans; and (5) energy-intensive farming systems.

6.2 Emerging problems in agriculture

Green revolution contributed to food security through increased food production and reduced volatility of food grain prices; and also demonstrated that agricultural development provides an effective means for accelerating economic growth and reducing poverty. But, post-green revolution input-intensive conventional agriculture production systems have led to several concerns, such as:

- Declining factor productivity
- Declining ground water table
- Development of salinity hazards
- Deterioration in soil fertility
- Deterioration in soil physical environment
- Biotic interferences and declining biodiversity
- Reduced availability of protective foods
- Air and ground water pollution
- Stagnating farm incomes

Therefore, the current state of agricultural production management systems is posing a threat to food security and livelihood of farmers, especially to resource poor and under-privileged small farm holders in vulnerable ecologies. Hence, the agronomic management in conventional crop production systems needs to be looked into critically and understood with an overall strategy of: (i) producing more food with reduced risks and costs; (ii) increasing input use efficiency, viz. land, water, nutrients, pesticides and labour; (iii) improving and sustaining the quality of natural resources; and (iv) mitigating emissions and greater resilience to changing climates.

6.3 Conservation agriculture: a new paradigm in crop production

Widespread resource degradation problems under conventional agricultural production system, and the need of reducing production cost, increasing profitability and making agriculture more competitive, made the conservation issues more imperative. Globally innovations of conservation agriculture-based crop management technologies are said to be more efficient, use less input, improve production and income, and address the emerging problems (Gupta and Seth, 2007). Additionally, secondary drivers as (i) availability of developed farm machinery; (ii) availability of new biocide molecules for efficient insect, pest disease and weed control; (iii) ever decreasing labour force and ever increasing labour cost; (iv) increasing production cost, energy shortage, erosion losses, pollution hazards and escalating fuel cost; and (v) residue burning have accelerated change in thinking of researchers, policy makers and farmers to adopt modified methods for cultivation of crops aimed at improving productivity and resource-use efficiency (Jat *et al.*, 2011)

Conservation agriculture (CA) practices are designed to achieve agricultural sustainability by implementation of sustainable agricultural management practices that minimize environmental degradation and conserve natural resources while maintaining higher yield, profitable systems and also improve the biological functions of the agro-ecosystem with limited mechanical farm practices and judicious use of external and chemical inputs. It is characterized by three linked principles, viz. (i) minimum mechanical soil disturbance, (ii) permanent organic soil cover and (iii) diversification of crop species grown in sequences and/or associations. A host of benefits can be achieved through employing components of conservation agriculture, including reduced soil erosion and water runoff, increased productivity through improved soil quality, increased water availability, increased biotic diversity, and reduced labour demands.

Tillage is an age-old practice of crop production. Jethro Tull (1671-1741), known as 'Father of Tillage' came up with a book entitled "Horse Hoeing Husbandry" where he propounded a theory that 'Soil particulars are ingested through openings in plant roots due to the processes caused by the swelling of growing roots'. He carried out numerous field experiments dealing with cultural practices leading to development of drill and horse drawn cultivator and advocated that soil should be finely pulverized for better growth of crop plants. The vulnerability of plough-based agriculture was exposed during Dust Bowl era (1931-39); as wind blew away the precious top soil from the drought-ravaged southern plains of US, leaving behind the failed crops and farms. However, there was no answer to solve the question of soil degradation. Then what Nature magazine termed "an agricultural bombshell" was dropped by Faulkner on July 5, 1943; with the first issue of his masterpiece book `Plowman's Folly and A Second Look.' The author opened a new era by simply saying a key sentence "The fact is that no one has ever advanced a scientific reason for ploughing." His idea was in sharp contrast to that advocated by Jethro Tull, and blamed the then universally used mouldboard plough for disastrous tillage of the soil. He questioned the use of plough for cultivation of crops, and argued that all standard wisdom used as a rationale for ploughing the soil was invalid. His ideas were considered 'mad' and without merit, until after his death when soil experts and scientists began to admit "We didn't pay attention, and we should have." Time magazine called this concept "one of the most revolutionary ideas in agriculture history." He is regarded as one of the first true conservationists. Faulkner's idea received further momentum with the findings of Masanobu Fukuoka, as outlined in his book 'The one straw revolution.' In this book, Masanobu Fukuoka presented a radical challenge to the global systems that we can rely on for our food. At the same time, it is a spiritual memoir of a man whose innovative system of cultivating the earth reflects a deep faith in the wholeness and balance of the natural world. For more than 65 years Masanobu Fukuoka worked to develop a system of natural farming. He did not plough his fields, used no agricultural chemicals or inorganic fertilizers, did not flood his rice fields as farmers have done in Asia for centuries, and yet his yields equalled or surpassed the most productive farms in Japan.

6.4 Prospects of conservation agriculture

Conservation agriculture has increased crop yields compared with conventional tillage in many countries, viz. USA, Australia, Mexico, Canada and Brazil. For example, sizable yield increases and income stability have led to wide-scale adoption of CA among farming community in Brazil (Saturnino and Landers, 2001). Similarly farmers in developing countries like India have also started to practice CA technologies. For example, zero-till (ZT) wheat in the rice–wheat system is currently being practiced on >3 million ha in north-western parts of the Indo-Gangetic Plains. The major CA based technology being adopted in this region is zero-till (ZT) wheat in the rice-wheat system; and it is now foreshadowing nothing less than the end of an age-old concept, popularly known as "more you till and more you eat". Globally, about 155 million ha area is practiced following the concepts and technologies for conservation agriculture; the major countries being USA, Brazil, Argentina, Canada and Australia (Table 6.1). Adoption and spread of ZT wheat has been a success story in north-western parts of India due to:

(i) reduced cost of production,

(ii) enhanced soil quality, i.e. soil physical, chemical and biological conditions,

(iii) (increased C sequestration and built-up in soil organic matter,

(iv) reduced incidence of weeds,

(v) increased water and nutrient-use efficiencies,

(vi) increased system productivity,

(vii) advances in sowing date,

(viii) greater environmental sustainability,

(ix) increased residue breakdown with legumes in the rotation,

(x) reduced temperature variability, and

(xi) opportunities for crop diversification and intensification.

Table 6.1: Global adoption of conservation agriculture systems

Country	Area (M ha)	% of global area
USA	35.6	23.0
Brazil	31.8	20.5
Argentina	27.0	17.4
Canada	18.3	11.8
Australia	17.7	11.4
China	6.7	4.3
Russian Federation	4.5	2.9
Paraguay	3.0	1.9
Kazakhstan	2.0	1.3
Others	8.2	5.3
Total	154.8	100.0

Source: www.fao.org/ag/ca/6c.html dt. 01.12.2014

The number of tillage operations done under CT (4-6 operation) is much more than under CA (only 1 operation for sowing) thereby saving time and cost. Besides reducing the cost of tillage operation, CA also saves irrigation water and reduces CO emission. For example, wheat under ZT requires 30-50 % less water in first irrigation and 15-20% less water in subsequent irrigation resulting in a saving of about 36 % water under ZT. It was also noticed that there was about 70-75 % reduction in CO emission under ZT wheat than in 2 CT (Table 6.2).

Table 6.2: Environmental impact of ZT wheat in Haryana and Bihar

Particulars	Haryana		Bihar	
	ZT	CT	ZT	CT
Diesel consumption for tillage operation (L/ha)	10.0	39.1	10.0	34.3
Irrigation water use (m^3/ha)	1710	2150	932	1134
Total CO2 emission (kg/ha)	21.6	84.5	21.6	74.0

Source: Pal *et al.* (2010)

Conservation agriculture has to addresses the complete agricultural system—the 'basket' of conservation-related agricultural practices. Initially it was considered that the practices of minimal tillage, permanent residue cover and planned crop rotations are to be prescribed simultaneously for making the CA successful. However, later on it was advocated to include integrated weed management as another crucial component for successful implementation of CA; as weeds are one of the most difficult management issues within this system. For example, adoption of ZT wheat in the rice–wheat cropping system in Pakistan's Punjab province increased during the initial years of its introduction, but later it showed a significant decrease in the proportion of adoption due to higher weed infestation. About 39% ZT users of this region faced an increase in weed problems due to ZT, while 37% reported no effect and 24% mentioned a decrease in weed infestation (Tahir and Younas, 2004). It was noted that the ZT adopters, non-adopters, and disadopters differ significantly in terms of their resource base; and disadopters also had more problems in controlling weeds. Giller *et al.* (2009) reported that weeds are 'Achilles heel' as they can affect crop yields and sustainability of CA systems.

6.5 Weeds in CA system

Tillage operation affects weeds by uprooting, dismembering and burying them deep enough to prevent further emergence. Ploughing also moves weed seeds both vertically and horizontally, and changes the soil environment; thereby promoting or inhibiting germination and emergence of weed seeds. Under CA reduction in tillage intensity and frequency increases weed infestation. Compared to conventional tillage (CT), presence of weed seeds is more in the soil surface under ZT, which causes higher weed germination and infestation. An increase in weed infestation was observed in aerobic direct seeded rice than in conventionally puddled transplanted rice (Singh *et al.*, 2008). Similarly, Mishra *et al.* (2012a) recorded that a ZT–ZT sequence favoured higher weed growth over a CT–CT sequence in rice–wheat cropping system. While in the initial year weed growth was not higher under the ZT–ZT sequence, but in the third year of experiment total dry weight of weeds was significantly higher under the ZT–ZT than the CT–CT tillage sequence (Table 6.3).

Table 6.3: The effect of tillage sequence on total weed dry weight (g/m^3) at harvest in rice and wheat

Tillage sequence	Rice	Wheat
ZT– ZT	154*	130a
ZT– CT	177	114
CT– ZT	102	131
CT– CT	99	96

*Data within a column with same letter do not differ significantly (P= 0.05).
Source: Mishra *et al.*, 2012a

6.6 Weed management

Management of weeds is a major issue in agricultural production system, particularly under CA where the infestation is likely to be higher than conventional intensive–tillage. Understanding ecology, seed bank and dynamics of specific weed flora is essential for developing effective management strategies in divergent situations. Weed control in CA is a greater challenge than in conventional agriculture because there is no weed seed burial by tillage operations. The behaviour of weeds and their interaction with crops under CA is complex and not fully understood. The weed species that germinate in response to light are more problematic in CA. In addition, perennial weeds are also more challenging to control in this system. In the past, attempts to implement CA have often resulted in a yield penalty because reduced tillage failed to control weed interference. However, the recent development of post-emergence broad spectrum herbicides provides an opportunity to control weeds in CA. Crop yields can be similar for conventional and conservation tillage systems if weeds are controlled and crop stands are uniform. Various approaches employed to manage weeds successfully in CA systems includes preventive measures, cultural practices (tillage, crop residue as mulches, intercropping, cover cropping, competitive crop cultivars), use of herbicide-tolerant cultivars, and herbicides.

a. Preventive measures

Preventive weed control encompasses all measures taken to prevent or arrest the introduction and arrest of weeds. Weed seeds resembling the shape and size of crop seeds are often the major source of contamination in crop seeds. Contamination usually occurs at crop harvesting if the life cycle of crop and weeds is of similar duration. Preventive measures are the first and most important steps to manage weeds, in general and especially under CA, as the presence of even a small quantity of weed seeds may cause a serious infestation in the forthcoming seasons. The various preventive measures include the following:

- Use weed-free crop seed
- Prevent the dissemination of weed seeds/propagules from one area to another or from one crop to another by using clean machinery/implements, screens to filter irrigation water and restricting livestock movement
- Use well-decomposed manure/compost so that it does contain any viable weed seeds
- Remove weeds near irrigation ditches, fence rows, rights-of-way, etc. prior to seed setting

- Mechanically cut the reproductive part of weeds prior to seed rain
- Implement stringent weed quarantine laws to prevent the entry of alien invasive and obnoxious weed seeds/propagules into the country.

b. Cultural practices

A long-term goal of successful and sustainable weed management is not to control weeds in a crop field, but also to create a system that reduces weed establishment and minimizes weed competition with crops. Further, since environmental protection is a global concern, the age-old weed management practices, viz. tillage, mulching, intercropping, cover crops, crop rotation/ diversification and other agro-techniques once labelled as uneconomical or impractical should be relooked and given due emphasis in managing weeds under CA. One of the important pillars of CA is ground cover with dead or live mulch, which provides less time for weeds to establish during fallow or turnaround period. Some other common problems under CA include emergence of new weeds from recently produced weed seeds that remain near the soil surface, lack of disruption of perennial weed roots, interception of herbicides by thick surface residue, and change in the timing of weed emergence. Shrestha *et al.* (2002) concluded that long-term changes in weed infestation were driven by the interaction of several factors including tillage, crop rotation, crop type, environment and timing and type of weed management practices.

c. Tillage

Tillage is an essential component of conventional agricultural system and it is the most important among the traditional means of weed management practices in agriculture. The effect of primary tillage operation on weed management mainly depends on the type of implement used and tillage depth. These factors impact the weed seed and propagule distribution over the soil profile; and therefore directly affect the number of weeds that can emerge in a field. Differential distribution of seeds in the soil profile subsequently leads to changes in weed population dynamics. Weed seeds buried deep germinate but fail to emerge due to the thick soil layer above it, resulting in death of the weed seedling. Tillage operation stimulates germination and emergence of many weed seeds through the exposure to sunlight. ZT wheat in a rice–wheat cropping system reduces infestation of canary grass (*Phalaris minor* Retze) which is highly competitive and can cause drastic wheat yield loss under heavy infestation and it also favours the infestation of toothed dock (*Rumex dentatus* L.) and cheese weed mallow (*Malva parviflora* L.) and wild oat. Cheese weed mallow is favoured by shallow seed burial and scarification leading to more weed population under a ZT system.

A reduction in weed density can be seen if the depletion of weed seed bank is greater than weed seed shedding. However, this situation can rarely be achieved with zero tillage. Therefore, weed density in zero tillage system is generally higher than in plough-based system. The findings of a long-term field experiment with four tillage systems adopted for 12 consecutive years in a continuous wheat or a pigeon pea-wheat rotation showed that total weed seedling density in ZT, minimum tillage using rotary harrow (15 cm depth), and chisel ploughing (45 cm depth) was relatively higher in the 0–15, 15–30, and 30–45 cm soil layers, respectively (Barberi and Lo Cascio, 2001). But ZT may affect seedling emergence of some weed species under a particular cropping system.

d. Stale seedbed

Seedbed preparation can contribute to weed management by affecting weed seed dynamics and seedling densities at planting. In CT, ploughing at a regular intervals controls initial weed population before crop sowing. Seedbed preparation affects the weed infestation in two ways: (i) it destroys the emerged weed plants after primary tillage, and (ii) it stimulates germination and emergence of weed seeds and reallocation of seeds towards the soil surface; this phenomena could be exploited to manage weeds through application of the stale seedbed technique.

No-till stale seedbed practice can help to reduce weed pressure in CA systems. In this technique, the field is irrigated 10–15 days prior to actual seeding to favour the germination of weed seeds lying on the soil surface. Emerged weeds are then destroyed by the application of non-selective herbicides like glyphosate, paraquat or ammonium glufosinate. It depletes the seed bank in the surface layer of the soil and reduces subsequent weed emergence. Where light rains occur for an extended period before the onset of the monsoon or irrigation is available, it may be possible to kill several flushes of weed growth before planting. To ensure success, cropping should be delayed until the main flush of emergence has passed. However, this practice may not be exploited where the season available for crop growth is short, which may reduce the yield potential of the crop. The main advantage of the stale seedbed practice is that the crop emerges in a weed-free environment, with a competitive advantage over late-emerging weed seedlings. The practice of false seed bed technique may decrease weed infestation in crops by 80% or more compared to standard seedbed preparation.

In many developed countries the stale seedbed technique is widely used to manage weedy rice and barnyard grass in rainfed rice. Stale seedbeds reduce weed population in direct-seeded rice and are especially more effective when combined with notill practice. Pittelkow *et al.* (2012) reported that ZT stale seedbed practice was effective at reducing the population of sedges and grasses, but not for controlling redstem weeds. This practice is very effective in ZT wheat in the north-western Indo-Gangetic Plains.

e. Crop residues

Crop residues present on the soil surface inhibit germination and emergence of weed seeds by interfering with sunlight availability and creating physical impedance, as well as improve soil and moisture conservation, and soil tilth. Residues on the soil surface can vary greatly in dimension, structure, distribution pattern and spatial heterogeneity. Weed biology, and the quantity, position and allelopathic potential of the crop residues may influence weed germination. Soil cover using crop residues is a useful technique to manage weeds. Weed emergence generally declines with increasing residue amounts. However, the emergence of certain weed species is also favoured by some crop residue at low amounts. For example, germination and growth of wild oat and animated oat (*Avena sterilis* L.) may get stimulated with low levels of wheat residue. High amounts of crop residues have implications for weed management in CA through reduced and delayed weed emergence. The crop gets competitive advantage over weeds due to delayed weed emergence, which results in relatively less impact on crop yield loss. Further, late emerging weeds produce less number of seeds than the early emerging ones (Chauhan and Johnson, 2010). For example, the residue of Russian vetch (*Vicia villosa* Roth) and rye (*Secale cereal* L.) reduced total weed density by more than 75% compared with the treatments with no residue (Mohler and Teasdale, 1993). The presence of rye mulch in corn significantly reduced the emergence of white lamb quarter, hairy crabgrass (*Digitaria sanguinalis* (L.) Scap.), and common purslane (*Portulaca oleracea* L.) and total weed biomass (Mohler and Calloway, 1992). However, crop residues alone may not be able to fully control weeds, residence suppressed weeds early in the growing season but herbicide was needed to achieve season-long weed control. The effectiveness of crop residue to reduce weed emergence also depends upon the nature of weed species to be controlled. Chauhan and Abugho (2012) reported that 6 t/ha crop residues reduced the emergence of jungle rice, crowfoot grass and rice flat sedge by 80–95 % but only reduce the emergence of barnyard grass by up to 35 %.

f. Intercropping

Intercropping involves growing a smother crop between rows of the main crop such that the competition for water or nutrients does not occur. Intercrops help to effectively pre-empt resources used by weeds and suppress weed growth, and hence can be used as an effective weed control strategy in CA. for example, alfalfa + barley, alfalfa + oats, fababean + red clover, maize + Italian ryegrass/ perennial ryegrass, maize + redclover, maize/cassava + cowpea/peanut/sweet potato, pigeonpea + urdbean/mungbean/soybean/cowpea/sorghum, rice + azolla, sorghum + cowpea/mungbean/peanut/soybean, chickpea + mustard etc. are some successful weed suppressing intercropping systems.

Intercropping of short-duration, quick-growing, and early-maturing legume crops with long-duration and wide-spaced crops leads to quickly ground cover, with higher total weed suppressing ability than sole cropping. Intercropping controls weed by increasing shade through canopy cover and crop competition. Like cover crops, intercrops also increase the ecological diversity in the crop field. In addition, they often compete better with weeds for light, water and nutrients. Success of intercropping depends on the best match between the requirements of the component crops for light, water and nutrients. Many short duration pulses like soybean, greengram and cowpea effectively supress weed infestation without reducing the yield of the main crop. For instance, total weed growth reduced under intercropping combinations of chickpea+mustard over the sole chickpea crop without losing productivity of the main crop (Rathi *et al.*, 2007). Similar observations were also recorded by Dubey (2008) under a maize+cowpea intercropping system. Compared with the sole crop, increased canopy cover and decreased light availability for weeds in maize–legume intercropping was responsible for the reduction in weed density and dry matter (Kumar *et al.*, 2010). However, intercropping cowpea in maize under CA had the greatest impact on weeding activities in the farmer's field, with labour hours increasing by 40 % due to the additional precision required for weeding compared with maize-only fields (Lai *et al.*, 2012).

g. Cover cropping

Ground cover with dead or live mulch, allowing less time for weeds to establish during fallow or turnaround period is an important component of CA technology. Inclusion of cover crops in a rotation between two main crops is a good preventive measure when developing a weed management strategy. Cover crops are fundamental and sustainable tools to manage weeds, optimize the use of natural resources, and reduce water runoff, nutrient leaching and soil erosion. Competition from a strong cover crop can virtually shut down the growth of many annual weeds emerging from seeds. Aggressive cover crops can substantially reduce the growth and reproduction of perennial weeds that regenerate from roots, rhizomes or tubers, and are more difficult to control. Cover crop's effect on weeds largely depends upon the type of species and weed community composition. Weed suppression is exerted partly through competition for light, water and nutrients during the growing cycle of cover crop and partly through physical and chemical effects that occur when cover crop residues are left on the soil surface as dead mulch or ploughed down.

Weed pressure in CA can be reduced by including short-duration legume crops e.g. green gram, cowpea, sesbania etc., during the fallow period between harvesting wheat and planting rice. This practice facilitates emergence of weeds during the legume period (stale seedbed effects) and reduces the population during the rice season (Kumar *et al*., 2012). The density of annual ryegrass plants in a wheat crop decreased to one third after green manured lupins compared with the harvested lupin crop.

Growing green manure or cover crops in the summer season or as a relay crop to efficiently suppress weed growth is a cost and labour efficient practice. Therefore, green manures are sometimes also called the herbicides of small farmers. Perennial grasses such as cogon grass (*Imperata cylindrical* (L.) P. Beauv.), bermuda grass (*Cynodon dactylon* (L.) Pers.), and other problem weeds like *Striga* spp. and Siam weed (*Chromolaena odorata* (L.) can be suppressed by one or two seasons of cover crops. In CA, a number of cover crops including legumes (sesbania, sunhemp, soybean, cowpea, alfalfa and clover) and non-legumes (rapeseed, rye, sunflower, buckwheat and sudan grass) can be cultivated to suppress various weeds.

h. Crop diversification

Crop rotation refers to alternating different crops in a systematic sequence on the same piece of land. It limits the weed population and prevents shift of weeds as the weed species tend to thrive in a crop with similar growth requirements. Different crops require different types of cultural practices, which disrupt the growing cycle of weeds and prevents any weed species to dominate. For example, Johnson grass generally becomes a predominant weed in a continuous maize system but may be controlled by rotating with cotton. In mono-cropping systems, several weed species persist and expand rapidly. Cropping sequences provide varying patterns of resource competition, allelopathic interference, soil disturbance and mechanical damage, and thus provide an unstable environment that prevents the proliferation and dominance of a particular weed, and discourages growth and reproduction of troublesome weed species. The prolonged cultivation of the rice–wheat system in north-western India has resulted in increased population of sedges and grassy weeds. The diversification of the system even for a short period of time and intensification by including summer legumes or green manuring crops decreases the weed emergence (Singh *et al.,* 2008) (Table 6.4).

Table 6.4: Density of different weed species (no./m^3) and weed dry matter production (g/m^3) in rice at 25 DAT under different crop sequences and grain yield

Crop sequence	Grasses	Sedges	Broad-leaved weeds	Weed dry matter	Rice yield (t/ha)
Rice–wheat	31.7	23.2	22.4	28.0	3.81
Rice–chickpea	32.3	20.3	13.5	25.3	3.88
Rice–wheat–greengram	9.9	7.3	3.3	9.9	4.09
Rice–wheat–Sesbania (GM)	17.4	13.0	15.1	19.4	4.21
Rice–mustard–greengram	16.2	9.9	7.2	15.9	4.05
Rice–lentil–cowpea (F)	15.7	8.4	4.7	13.3	4.13
Rice–pea	32.2	20.7	16.7	26.4	4.01
Rice–lentil+mustard (3:1)–cowpea (F)	17.2	11.5	6.5	15.1	4.15
Rice–maize+pea(1:1)–cowpea (F)	19.4	12.6	11.1	20.0	4.18
Rice–potato–green gram	12.5	10.4	7.6	14.5	4.19

Source: Singh *et al*., 2008

i. Cultivar competitiveness

Crop species and cultivars differ in their competitiveness with weeds. The expression of competitive advantage of crop genotypes against weeds is strongly influenced by environmental conditions. The competitive ability of a crop variety is reflected either by its ability to reduce weed growth and seed production or to tolerate weed interference and maintain higher levels of grain yield. Different genotypes of the same crop may differ in their competitive ability against weeds due to varying morphological traits. Although there is conflicting evidence as to which crop characteristics contribute most to competitiveness, several studies have highlighted the role of rapid germination and emergence, vigorous seedling growth, rapid leaf expansion, rapid canopy development, extensive root systems, and also production of allelopathic compounds by the crop. However, mostly the crop competitiveness is enhanced by vigorous growth that reduces light quality and quantity beneath the crop canopy.

j. Planting geometry

Planting density and pattern modify the crop canopy structure, and in turn influence weed smothering ability. Narrow row spacing brings variation in microclimate, viz. light intensity, evaporation and temperature at soil surface. The establishment of a crop with a more uniform and dense plant distribution results in better use of light and water, and leads to greater crop competitive ability. Crops grown in narrow rows start competing with weeds at an earlier stage than those in wide rows because of more rapid canopy closure and better root distribution. Narrow row widths and a higher seeding density will reduce the biomass of late-emerging weeds by reducing the amount of light available for weeds located below the crop canopy. Reduced growth of weeds was reported due to increased population and decreased spacing in rice. The LAI of closely-planted rice increased but PAR decreased, and grain yield was significantly higher than the widely-spaced crop. Similarly, bidirectional sowing and closer row spacing are quite effective in suppressing the growth of little seeded canary grass in wheat.

k. Sowing time

Planting time influences the occurrence and manifestation of weed species. Thus, sowing time should be manipulated in such a way that ecological condition for the germination of weed seeds are not met. In the north-western part of the Indo-Gangetic Plains, farmers advance wheat seeding by 2 weeks to get a head start over the noxious weed little seed canary grass and provide higher yield (Singh *et al.*, 1999).

l. Farm machinery

For conservation agriculture suitable machine-driven farm machineries are required for land levelling, sowing, fertilizers and irrigation application, weeding, harvesting and other farm operations. Hence, the availability of farm machineries is a paramount importance for adoption of CA technology by farmers. For example, Farooq *et al.* (2007) noticed that access to ZT drills contributed towards the adoption pattern of the ZT wheat technology in Pakistan's Punjab province. `No-till' seed drill has been invented by Morton C. It has allowed the farmers to sow seed without tilling the land. Direct drilling with ZT drill is a practice that addresses the issues of labour, energy, water, soil health, etc. However, this machine faces some difficulties if the quantity of crop stubbles is high and this situation commonly occurs in CA systems. Harvesting of rice, wheat and many other crops like maize, sorghum, pegionpea, chickpea, greengram, blackgram, etc. is being done through combine harvester in many parts of India. In most regions, the crop residue lying on the soil surface is burnt so as to prepare the field for sowing of next crop. This is a very unscientific practice as it leads to environmental problem, loss of C and other essential nutrients. Despite some measures taken by different states and ban imposed by Supreme Court, burning is still the most common means of straw disposal.

m. Chemical weed management

The use of herbicides for managing weeds is becoming popular because they are cheaper than traditional weeding methods, require less labour, tackle difficult-to-control weeds, and allow flexibility in weed management. Herbicides are an integral part of weed management in CA systems. However, in CA systems, herbicide rotation or integration of weed management practices is preferred over continuous use of a single herbicide over a long period of time which may result in the development of resistant biotypes and negative effects on the succeeding crop and environment. In CA, the diverse weed flora that emerges in the field after harvesting the preceding crop must be killed using non-selective herbicides like glyphosate, paraquat and ammonium-glufosinate. Non-selective burn-down herbicides can be applied before or after crop planting but prior to crop emergence in order to minimize further weed emergence.

Unlike in conventional agriculture system, crop residues present at the time of herbicide application in CA system decreases the effectiveness of herbicide as the residues intercept applied herbicide droplets and reduces the amount of herbicide that reaches the soil surface. Proper selection of herbicide formulation for application under CA systems is necessary to increase their efficacy. For example, pre-emergence herbicides applied as granules may provide better weed control than liquid-forms in no-till systems. Some herbicides intercepted by crop residues in CA systems are prone to volatilization, photo-degradation and other losses. The extent of loss, however, varies depending upon chemical properties and formulations. Herbicides with high vapour pressure are susceptible to volatilization from the soil surface. Climatic conditions and herbicide application methods significantly affect herbicide persistence under CA systems. Crop residue can intercept almost 15–80% of the applied herbicide which results in reduction of herbicide efficacy in CA systems (Chauhan *et al.*, 2012). Weeds control by the application of herbicides was better in the CT system (80–96%) than in the ZT system (50–61%) (Chauhan and Opena, 2012). Choosing an appropriate herbicide and timing of its application is critical in CA systems as weed control under no-till systems varies with weed species and herbicides used. Pre-emergence herbicides may not be as efficient in controlling weeds in CA systems due to the presence of crop residues which can bind to soil-applied herbicides and favour the weed seedlings to escape the applied herbicides.

6.7 Conclusion

CA is a complex suite of resource-efficient technologies. It is possible to achieve the same or even higher yields with conservation agriculture compared with conventional agriculture. Altering tillage practices changes the depth of weed seeds in the soil, which shifts weed species and affects the efficacy of control practices. ZT systems cause a shift in weed flora, and may result in emergence of perennial weeds like purple nut sedge, Bermuda grass and Johnson grass in most crops; and others like cheese weed mallow and toothed dock in wheat. Restricting tillage also reduces weed control options and increases reliance on herbicides; consequently, evolvement of weed resistance to herbicides has become a serious and escalating problem for many CA farmers worldwide. The use of HT crops further aggravates the situation. ZT along with residue has beneficial effects on soil moisture, temperature moderation and weed control. CA is a machine-, herbicide- and management driven agriculture for its successful adoption. Integrated weed management involving chemical and non-chemical methods (residue, cover crops, varieties etc.) is essential for success of CA systems in the long term.

References

Chauhan BS, Opena J (2012) Effect of tillage systems and herbicides on weed emergence, weed growth, and grain yield in dry-seeded rice systems. Field Crops Research 137: 56-69.

Chauhan BS, Singh RG, Mahajan G (2012) Ecology and management of weeds under conservation agriculture: A review. Crop Protection 38: 57-65.

Dubey RP (2008) Effect of weed control and nitrogen application rates on weed infestation and productivity in maize– cowpea intercropping system. Indian Journal of Weed Science 40: 155- 158.

Farooq U, Sharif M, Erenstein O (2007) Adoption and impacts of zero-tillage in the rice-wheat zone of irrigated Punjab, Pakistan. New Delhi: CIMMYT and the Rice-Wheat Consortium for the Indo-Gangetic Plains, pp. 9–18.

Lai C, Chan C, Halbrendt J, Shariq L, Roul P, Idol T, Ray C, Evensen C (2012) Comparative Economic and Gender, Labor Analysis of Conservation Agriculture Practices in Tribal Villages in India. International Food and Agribusiness Management Review 15: 73-86.

Mishra JS, Singh VP (2012a) Tillage and weed control effects on productivity of a dry seeded rice– wheat system on a vertisol in central India. Soil and Tillage Research 123: 11–20.

Pal S, Sekar I, Kar A (2010) Environmental impact of resource conservation technology: the case of zero-tillage in the rice-wheat system of South Asia. Report to the CGIAR standing panel on impact assessment.

Singh RK, Bohra JS, Srivastava VK, Singh RP (2008) Effect of diversification of rice– wheat system on weed dynamics in Rice.Indian Journal of Weed Science 40: 128-131.

Singh S, Kirkwood RC, Marshall G (1999) A review of the biology and control of Phalaris minor Retz. (little seed canary grass) in cereals. Crop Protection 18: 1-16.

7

In-Situ Soil Moisture Conservation Techniques

7.1 What is Soil Moisture Conservation Technique?

In situ soil moisture conservation means capturing rain water and retaining it in soil for utilization by plants for better growth and increase in biomass and grain yield. This is achieved through rain water harvesting on farm or field where crops are planted to get benefit from the conserved rainwater. It reduces rainwater loss through runoff and increases rain water infiltration time, thereby increasing the amount of rain water in the soil that will be useful for crop production.

7.2 Implementation of in-situ soil moisture conservation techniques

Storage of rainfall in soil at the place where it falls is termed as "in situ" soil moisture conservation. To increase the availability of soil moisture to the agricultural crops, in the recent time it is necessary to adopt in-situ moisture conservation techniques in addition to the large scale soil moisture conservation and rain water harvesting structures in the watershed. The principles behind the recommendation of different practices are: i) to increase the infiltration by reducing the runoff rate, ii) temporarily pounding of rain water on the surface of the soil to increase the opportunity time for infiltration and iii) modifying the land configuration for water harvesting. Earlier efforts for moisture conservation were concentrated upon construction of various types of bunds across the land slope to control soil erosion and conserve soil. However, all the erosion control measures led to accumulation of water against the structures rather than its proper and uniform distribution in the inter-terraced area and at times led to reduction in crop yield. To overcome these problems in-situ moisture conservation techniques are recommended. In situ soil moisture conservation can be accomplished through:

- Cultural/agronomic methods
- Mechanical methods
- Agrostological/biological methods

Soil moisture storage from rainfall is influenced by many factors likes quantity and intensity of rainfall, soil properties such as texture, structure, depth, presence of subsoil hard pans, surface characters, slope, rate of infiltration and permeability, water holding capacity, vegetative cover etc.

7.3 Cultural /Agronomical methods

(i) **Addition of organic matter:** Addition of organic matter in the soil minimizes evaporation loss, improves physical properties and water holding capacity of the soil.

(ii) **Off season/summer tillage:** Furrows created in the field during ploughing can hold water in the depressions and thereby increase the infiltration of rain water. In the sloppy areas the plough furrows reduce the velocity of runoff water and improve storage. Summer tillage is a traditional tillage practice that helps in the storage of pre-sowing rainfall. When ploughing is done along contour, it is called as contour ploughing and is more helpful for in situ moisture conservation.

(iii) **Contour farming:** Ploughing the soil and sowing the crops along the contour reduce soil erosion and runoff. For example, Jowar sown in the black soil on contour line can restrict the runoff to 13.7% of the total rainfall and soil loss to 2.4 t/ha/year.

Figure 7.1: Contour farming

(iv) **Cover crops:** Soil erosion is reduced if the land surface is fully covered with foliage. Example: green gram, black gram, groundnut and fodder grasses like *Cenchrus glaucus*, *Cenchrus ciliaris*, dinanath grass, marvel grass etc. Both contour cropping and cover cropping can be practiced when the slope is less than 2 %.

Figure 7.2: Cover crop

(v) **Mixed cropping/ Inter cropping:** Cultivating a combination of crops with different planting times and different length of growth periods.

(vi) **Mulching:** Mulch is a layer of organic or inorganic material that is placed on the root zone soil of the plants. Examples of organic mulch materials include straw, wood chips, peat. Inorganic mulch in form of plastic sheeting is also used. Mulching is most suitable for low to medium rainfall areas and less suitable for areas with very high rainfall. Mulching is one of the most important agronomic practice that has direct influence on hydro-thermal regimes of soil by suppressing the weeds, reducing water evaporation, conserving the soil moisture, modifying the physical environment of soil and improving soil fertility.

Figure 7.3: Mulching in maize field

(vii) **Strip cropping:** Strip intercropping involves cultivation of erosion resistant crop and erosion permitting crop in alternate strips of 2-3 m width across the slope and along the contour. Erosion resistant crops include legumes and grasses with rapid canopy development. Example: cultivation of *Cenchrus glaucus* + *Stylosanthes hamata*.

(vii) **Conservation tillage:** In conservation tillage the tillage operations are limited to the single line where the crop is sown. In this practice the leftover stubbles of the previous crop is left on the field and the field is ploughed only at the sowing zone with minimum disturbance to the soil. Tillage should be reduced or eliminated (in severe cases) to maintain healthy organic matter in soil, to conserve the resources and reduce the evaporation loss by not disturbing the stubbles of the harvested crop. Conservation agriculture can improve the ability of the soil to absorb and retain water. Some studies mentioned that conservation tillage operations serve as an effective tool for retaining soil moisture content.

(ix) **Crop rotation:** Crop rotation is the practice of growing different types of crops successively on the same plot of land over a year. Incorporation of legumes in cereal crops can reduce soil erosion, restore soil fertility and conserve soil moisture. It minimizes insects and weeds infestation, optimizes soil nutrients and improves soil health. Rotating shallow-rooted crops and deep-rooted crops in cropping sequence can be more effective for utilizing the soil moisture of different depths. Some studies report that crop rotation can increase the moisture level in soil and yields in rainfed farming.

7.4 Mechanical methods

The basic principles of mechanical methods are:

- shaping the land surface manually or with implements in such a way as to reduce the velocity of runoff,
- to allow more time for rainfall water to stand on the soil surface, and
- to facilitate more infiltration of rainfall water into the soil.

Choice of any particular mechanical method under a given situation is dependent upon rainfall characters, soil type, crops, sowing method and land slope.

(i) **Basin listing:** Formation of small depressions or basins of 10-15 cm depth and 10-15 cm width in the field at regular intervals using an implement called basin lister. These small basins collect rainfall water and help in infiltration. It is usually done before sowing and suitable for all types of soil and crop.

(ii) **Bunding:** Formation of narrow based or broad based bunds across the slope at suitable intervals depending on the slope of field checks the free flow of runoff water, impounds the rainwater in the inter-bund space, increases its infiltration and improves soil moisture storage. Leveling of inter-bund space is essential to ensure uniform spread of water and avoid water stagnation in patches.

It can be classified into three types:

(a) **Contour bunding:** Bunds of 1 m basal width, 50 cm top width and 50 cm height are formed along the contour to check runoff. The distance between two contour bunds depends on the slope. The inter bund surface is leveled and used for cropping. It is suitable for deep red soils with the slope less than 1%. It is not suitable for heavy black soils where bund tends to develop cracks on drying. Contour bunds are permanent structures and require technical assistance and heavy initial investment.

Figure 7.4: Contour bunding

(b) **Graded/field bunding:** Bunds of 30-45 cm basal width and 15-20 cm height are formed across the slope at 20-30 m interval depending on the slope to check runoff. The inter-bund area is leveled and used for cropping. It is suitable for medium deep to deep red soils with slope up to 1%. It is not suitable for black soils due to susceptibility to crack on drying. Bunds can be maintained for 2-3 seasons with reshaping.

Figure 7.5: Field bunding

(c) **Compartmental bunding:** Small bunds of 15 cm width and 15 cm height are formed in both directions (along and across the slope) to divide the field into small basins of 40 sq m size (8 × 5 m). It is suitable for red soil and black soil with a slope of 0.5-1%. The bunds are formed before the sowing or immediately after the sowing with local wooden plough. It is mostly suitable for the crops which are broadcasted. CRIDA has recommended this method as the best in situ soil moisture conservation measure for Tamil Nadu. Maize, sunflower and sorghum perform well in this type of bunding.

(iii) **Ridges and furrows:** Furrows of 30-45 cm width and 15-20 cm height are formed across the slope. The furrows guide runoff water safely when rainfall intensity is high and thus helps to avoid water stagnation. They also collect and store rainwater water when rainfall intensity is less. It is suitable for medium deep to deep black soil and deep red soil. It can be practiced in wide row spaced crops like chili, tomato, maize, cotton etc. It is not suitable for shallow red soil, shallow black soil and sandy soil. It is not suitable for broadcast sown crops and for crops sown at closer row spacing less than 30 cm. As furrows are formed usually before sowing, sowing by dibbling or planting alone is possible. Tie ridging is a modification of ridge and furrow method where the ridges are tied by a small bund at 2-3 m interval along the furrows. Random tie ridging is another modification where discontinuous furrows of 20-25 cm width, 45-60 cm length and 15 cm depth are formed between clumps or hills of the crops at the time of weeding. Another modification of ridge and furrow method is the sowing of crops in lines on flat beds and formation of furrows between crop rows at 25-30 DAS. This enables sowing behind plough or through seed drill.

Figure 7.6: Ridges and furrows

(iv) Broad Bed Furrow (BBF): Beds of 1.5 m width and 15 cm height with convenient length are formed which are separated by furrows of 30 cm width and 15 cm depth. Crops are sown on the beds at required intervals. It is suitable for heavy black soil and deep red soil. Broad bed furrow has many advantages over other methods, like,

- It can accommodate a wide range of crop geometry i.e., close as well as wide row spacing.
- It is suitable for both sole cropping and intercropping systems.
- Furrows serve to safely guide runoff water in the early part of rainy season and store rainwater in the later stages.
- Sowing can be done with seed drills.
- It can be formed by bullock drawn or tractor drawn implements. Bed former cum seed drill enables BBF formation and sowing simultaneously, thus reducing the delay between rainfall receipts and sowing.

(v) Dead furrow: At the time of sowing or immediately after sowing, deep furrows of 20 cm depth are formed at intervals of 6-8 rows of crops. No crop is raised in the furrows. Sowing and furrowing are done across the slope. It can be done with wooden plough in both black and red soils.

Figure 7.7: Dead furrow

7.5 Agrostological methods

Cultivation of grasses to reduce run off, control soil erosion and improve soil moisture storage called agrostological method. Grasses with their close canopy cover the soil surface and their profuse root system binds soil particles, provides excellent protection against runoff and erosion. The following are the various agrostological methods of in situ moisture conservation.

(i) **Pastures/grass lands:** Raising perennial grasses to establish grass lands is recommended for shallow gravelly, eroded, degraded soils. Grass canopy reduces splash erosion by intercepting rainfall, checks runoff and improves soil moisture storage.

(ii) **Strip cropping with grasses:** Alternate strips of grasses and annual field crops across the slope check runoff, erosion and increase moisture storage in soil.

(iii) **Ley farming:** Growing of fodder grasses or legume crops and annual crops in rotation is called ley cropping. Grasses and legumes like cenchrus, stylo are grown for 3-5 years followed by annual crops like sorghum for 2 year. Grasses or legume crops improve soil moisture conservation.

(iv) **Vegetative barriers:** One or two rows of perennial grasses are grown at suitable interval across the slope and along the contour. It serves as a block to free runoff and soil transport. Vetiver, cenchrus are suitable grasses. Plough furrows are opened with disc plough before the monsoon. 5–8 cm deep holes are formed at 20 cm interval and two slips per hole are planted in the beginning of rainy season. If fodder grasses like *Cenchrus glaucus* or marvel grass are used, fodder can also be harvested for animal feed. Vegetative barriers are best suitable for black soil. Unlike contour bunding, which develop cracks in black soils during summer, vegetative barriers do not allow such phenomenon in black soil. Hence, the vegetative barriers can be effectively maintained in black soil for 4-5 years.

7.6 Benefits of soil moisture conservation

Soil moisture conservation practices play a major role in crop production under dry land and rainfed farming systems. Proper crop management practices along with adequate water supply can improve the yields in rainfed crops. In rainfall deficit areas by applying proper soil moisture conservation technique the *rabi* crops can be grown successfully with residue soil moisture from the previous *kharif* crop. Depending on the moisture conserving method used, conservation techniques can also have other beneficial effects like weed management, protection of the soil surface from the inadequate consequences of heavy rain, runoffs, soil erosion and increased nutrient distribution in the soil. These methods can further involve in reducing water requirement by improving water use efficiency of the crops. It reduces waste management requirements by recycling organic waste and using plant and agricultural residues through decomposition in the soil.

7.7 Barriers of soil moisture conservation

Crop wastes are sometimes utilized as animal feed in some areas, providing extra investment for soil protection. Soil moisture conservation measures may not be used in all regions since growing new crops via mulch or other crop wastes may be challenging in non-mechanized agriculture. The burning of crop residues in the field is a barrier to use crop residues for in-situ soil moisture conservation. Some of the moisture conservation techniques like plastic mulching and strip cropping cannot applicable for all the crops. The techniques which involved high mechanization are not cost effective for small farmers. Lack of awareness about conservation agriculture practices among many farmers is also a major barrier for implementing moisture conservation practices.

7.8 Conclusion

Soil moisture conservation techniques are necessary to maintain optimum moisture level in the soil for supplying the required amount of water and essential nutrients to the crop plants and maintain physical, chemical and biological properties of the soil. Different techniques are followed in different regions. The choice of appropriate soil moisture conservation technique depends on soil, topography, climate and also on the farmer's choice. These techniques also help in attaining agriculture sustainability in dry land and rainfed farming systems. However, all the soil moisture conservation techniques have some advantages and drawbacks. According to some experimental findings, the combination of different soil moisture conservation techniques is a better approach than following a single technique.

References

Meena LR, Singh RK and Gautam RC. (2002). Effect of moisture conservation practices, phosphorus levels and bacterial inoculation on chickpea (Cicer arietinum) under rainfed condition. Indian Journal of Agronomy 47(3): 398–404.

Mishra JS, Singh VP and Yaduraju NT. (2005). Effect of tillage practices and herbicides on weed dynamics and yield of wheat (Triticum aestivum) under transplanted rice (Oriza sativa)–wheat system in Vertisols. Indian Journal of Agronomy 50(2): 106–109.

Phogat VK and Dharam Pal. (2005). Effect of deep tillage and gypsum on yield, and N, P, K and S uptake by mustard. Journal of the Indian Society of Soil Science 53(1): 134–136.

Singh Pratap, Mittal SP and Agnihotri Y. (1996). Moisture use efficiency and yield of mustard as affected by pre-sowing irrigation and mulch. Indian Journal of Soil Conservation 24(2): 128–131.

Singh RK, Prasad SN, Ali Shakil, Kumar Ashok, Singh KD, Prasad Arjun, Singh SV and Parandiyal AK. (2005). Onfarm evaluation of conservation measures to performance of rainfed crops in semi-arid region. Indian Journal of Soil Conservation 33(2): 141–143.

Vaidya VB, Varshneya MC, Bote NL and Naidu TRV. (1995). Estimation of thermal efficiency and apparent reflectivity of mulches using soil temperature. Journal of Maharashtra Agricultural Universities 20(3): 341–344.

8

Conservation Agriculture under Climate Change: Adaptation and Mitigation in India

8.1 Introduction

Conservation agriculture (CA) is an approach to regenerative agriculture and land management practices based on three interlinked principles: (1) continuous minimum mechanical soil disturbance using no-tillage or reduced tillage based crop cultivation, (2) maintenance of permanent soil cover using crop residues and cover crops and (3) diversification of cropping system using economically, environmentally and socially adapted crop rotation including legume crops and cover crops coupled with other complementary agronomic management practices (FAO, 2014). Minimum soil disturbance provides several benefits such as maintaining optimum proportion of respiration gases at the root zone of plant, lesser oxidation of soil organic matter, porosity maintenance for water movement and limits the re-exposure of weed seeds to sunlight that restricts germination. It also allows soil microbes to produce stable soil aggregates for efficient infiltration that referred as "biological tillage". Soil cover is very much essential to protect the soil surface from deleterious impacts of exposure to rain and sunlight and to provide constant food supply to soil microorganisms and plant root. As a result, there is an improvement in soil biodiversity, microbial activity, soil aggregation and carbon sequestration. Diversified crop rotation offers a diverse "diet" to the soil microorganisms and also helps in recycling the soil nutrients by the cultivation of different crops. Inclusion of legume crops in cropping sequence and crop rotation helps in biological nitrogen (N_2) fixation, minimum pest infestation and enhancement of soil biodiversity. Construed more broadly, it also encompasses natural resource management to upsurge the synergies between crop production and ecosystem conservation as it includes diversified farm practices such as watershed management, agroforestry, livestock and fodder management, improved fallows etc. CA also deals with the climate related issues and plays a major role in climate change mitigation by reducing greenhouse gas (GHGs) emission or carbon sequestration, conservation of biodiversity and ecosystem.

The CA can sequester the atmospheric carbon dioxide (CO_2) into soil carbon at slower rate. Incorporation of crop residues and perennial woody biomass such as fruit trees, live fences, etc. in the field accelerates the C sequestration rates.

8.2 Conservation agriculture in different ecosystems

A. Conservation agriculture in irrigated ecosystems

During the past few decades, an alternative to tillage based conventional agriculture, CA with the elements of site specific conservation technologies with similar or higher yields, high profitability and high use efficiency of external production inputs and natural resources, has emerged as major sustainable intensification strategy and practiced over 200 Mha land globally. In South Asia, CA is relatively a new concept compared to leading CA adopting countries, but estimate indicates that a partial CA based system (at least one crop has no till with or without crop residue retention) is spread to over 2.5 Mha field in South Asia (Jat *et al.,* 2020). In India, systematic researches on CA were started during mid1990's in rice-wheat cropping system of Indo-Gangetic plain (IGP). Over the past two decades, a significant advancement has been made in terms of development and refinement of tillage operation, crop residue management, crop establishment, CA machinery and associated component technologies adapted to CA based management practices in major production systems of the region. Zero-till wheat has been adopted significantly on a large area in the rice-wheat system of IGP of South Asia, with positive impacts on wheat yield, profitability, resource use efficiency, improvement in soil health and reduction in greenhouse gases emission and air pollution. Research on CA jointly by CIMMYT-ICAR researchers using meta-analysis of large dataset across South Asia have generated evidence on the performance of CA on key parameters like protein content, yield profit, water saving, GHGs emission, soil health, adaptive capacity to climatic risks etc. which shows that CA not only has multiple benefits but also has potential to contribute to the United Nation Sustainable Development Goals through alleviating the multiple stresses including the emerging social issues currently faced by agriculture in South Asia (Jat *et al*., 2020). However, the response for different indicators varies under various management circumstances such as cropping systems, soil types etc.

A large number of studies have been carried out on various elements of CA or CA in different cropping systems and agro-ecologies of South Asia. For example, zero tillage with residue retention had an average yield advantage of 5.2 %, irrigation water saving of 9.8 %, reduction in cost of cultivation by 14.2 %, increase in net return by 27.5 % and also a reduction of 12-33 % in global warming potential with full CA adoption. Many results further suggest that CA and its component technologies give real benefits in the cereal based cropping systems of South Asia, for its potential for increasing net return and there are opportunities for improving technology to maximize the expected benefits. The results of several studies on various aspects of CA on the key performance indicators such as crop productivity, economic profitability, water saving and climate benefit are summarized in Table 8.1.

Table 8.1: Evidences of climate-smart agricultural practices in South Asia

Sl.No.	Climate-smart technology	Crop/ cropping System	Grain productivity (kg ha^{-1})	Water saving (%)	GHG mitigation (kg CO_2 eq ha^{-1})
1.	Laser land levelling	Rice-wheat; maize-wheat	350-645	15-20	400-600
2.	Zero tillage (ZT) without residue	Wheat	300-450	-	1000-1500
3	ZT with residue	Wheat	410-660	6-10	1500-2000
4.	Direct seeded rice (DSR)	Rice	±150-220	12-18	1000-2250
5.	Permanent beds	Maize-wheat	450-950	65-70	800-1000
		Maize-mustard	400-1120	70-85	-
		Pigeonpea-wheat	400-900	15-25	
6.	Site specific nutrient management	Rice-wheat; maize-wheat	350-650	-	500-1000
7.	Sub-surface drip irrigation	Rice-wheat	280-550	40-50	500-1500
		Maize-wheat	500-1020	70-85	-
8.	Legume inclusion (Mungbean)	Rice-wheat; maize-wheat	320-350	-	65-120
9.	Crop diversification	Wheat and mustard based system	400-1050	15-75	-
10.	Portfolios of climate smart practices (LLL + ZT + Mulch + SSNM)	Rice-wheat	110-220	15-20	-

B. Conservation agriculture in rainfed ecosystems

Rainfed ecosystem in India occupies about 55% of the total cultivable area and produces about 40% food in the country. The predominant rainfed states in India are Andhra Pradesh, Telangana, Maharashtra, Karnataka, Tamil Nadu, MP, Orissa, Rajasthan and others. Soils are degraded and multi-nutrient deficiency with poor in soil organic carbon (SOC) is common. As Indian agriculture is rainfall dependent, sustainability is directly associated with the amount and distribution of rainfall (Srinivasarao *et al*., 2011c, d). In the past few decades, decrease rainfall events are seen though total amount of rainfall is not velry much varied. This leads to increase in frequency of mid-season drought which leads to crop failure and overall uncertainty in the food production systems. Therefore, resource conservation technologies (RCTs) for water, soil, carbon and nutrient conservation are much focused in the recent past. Conservation agriculture in true sense with three principles i) no-tillage or minimum tillage, ii) introduction of legume in crop rotation and iii) crop residue cover provides multiple benefits of resource conservation, but implementation in real field situation is often difficult due to the need of in-situ rain water conservation, need of tillage operation for surface crusting management for optimum seed germination and introduction of legume crops with mono-cropping system. This is the reason why the RCTs are more practical and implementable depending upon rainfall, soil type and other ecosystem properties (Srinivasarao *et al*., 2011b, 2012a).

Experiment in rainfed Alfisols showed that introduction of legume crop horsegram (post-rainy season) in maize (*kharif* season) is possible when sufficient terminal rainfall is received during *kharif* season. Taking a second crop in predominantly mono cropped region is purely dependent on optimal monsoon distribution (Srinivasarao *et al*., 2016, 2019a) (Figure 8.1). Another important constraint of crop residue recycling in rainfed agriculture is due to lower biomass production and its competing uses for animal feed. If CA system which broadly dependent on proper rainfall distribution in crop growing season, can be implemented successfully, it improves system productivity, SOC status, reduce soil erosion and improves biological activity in soil.

Another important technology for resource conservation in rainfed dryland areas is land treatments coupled with intercropping system. Land treatments such as construction of furrows, ridge-furrow, semi-permanent raised beds etc. will allow conserving rain water which is critical for sustainable agriculture in rainfed system and they also reduce soil and nutrient loss. If land treatments and legume based intercropping are done together, they create ample opportunities for climate change adaptation and mitigation (Srinivasarao *et al*., 2014, 2017, 2019a). Promising intercropping system for different rainfed ecosystems are presented in Table 8.2.

Table 8.2: Efficient inter-cropping systems in rainfed drylands for resource conservation in India

Sl.No.	System	Target domain	Adoption rate
1.	Groundnut + pigeonpea (7:1)	Rayalaseema, Andhra Pradesh	70%
2.	Cotton + sorghum + pigeonpea + sorghum (6:1:2:1)	Vidarbha, Maharashtra	40%
3.	Maize + pigeonpea (1:1)	Odisha	40%
4.	Sorghum + pigeonpea (2:1)	Telangana	40%
5.	Fingermillet + pigeonpea (10:1)	South Karnataka	20%
6.	Pearlmillet + pigeonpea (2:1)	Solapur, Maharashtra	35%
7.	Maize –wheat + raya (2:1)	Kandi region of Punjab	38%
8.	Maize + blackgram (2:2)	Southern Rajasthan	63%
9.	Pearlmillet + castor (2:1)	North Gujarat	15%

Soils are poor in SOC and fertility. Improving SOC with integrated nutrient management (INM) practices is the better option which conserves more water by improving water retention capacity of the soil and also improves nutrient availability in rainfed drylands. Important strategies are required for implementing INM in rainfed areas to generate organic resources with legume cover crops with off-season rainfall or introduction of legume crops in fallow period. Various organic resources such as farm yard manure, crop residues, legume cover crops, green leaf manure (glyricidia) and other green manure crops are added to the soil in different rainfed ecosystems of India. These methods improve SOC in soil and contribute to climate change adaptation by protecting the crop during mid-season drought by improving soil moisture retention capacity (Srinivasarao *et al.*, 2012a, b, c, d, e f). Each ton of SOC stocked in soil improves productivity of rainfed crops such as groundnut, soybean, safflower, finger millet, sorghum, pearl millet, cluster bean, lentil and rainfed rice to the range of 40 to 160 kg/ha (Figure 8.1). However, in rainfed dry land ecosystem, limited options are available in terms of water availability and implementation of conservation technologies in holistic manner at village level.

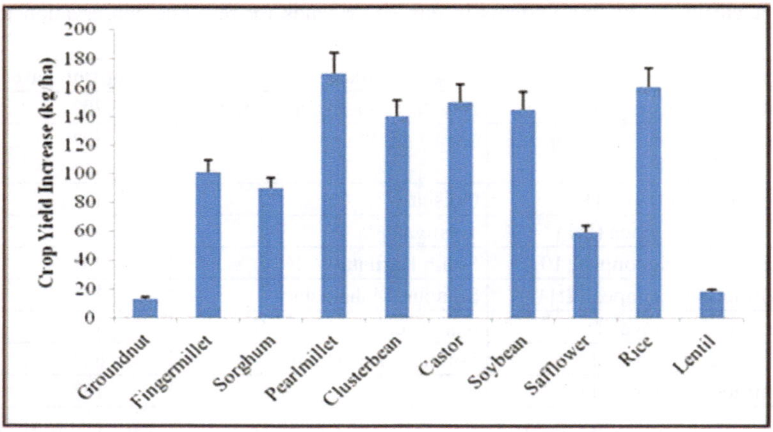

Figure 8.1: Increase in crop yield for every ton per hectare increase in soil organic C stock in the root zone in different rainfed production systems

8.3 Conservation agriculture for reducing GHG emission

Agriculture contributes to greenhouse effect through the emission of GHGs such as CO_2, CH_4 and N_2O. Rice fields submerged under water are the sources of CH_4 production. Nitrous oxide is produced in soil through nitrification and denitrification processes. The sources of CO_2 in agriculture are the soil management practices as tillage, use of fuel for various agricultural operations such and burning of crop residues. Manufacturing of farm implements, fertilizers and pesticides are the other sources of carbon dioxide. Resource conservation technologies have potential to enhance input use efficiency and mitigate GHG emission. The GHG mitigation as well as adaptive capacity of various RCTs to climatic risk is presented in Table 8.1 and a few of them are discussed below.

Zero tillage (ZT) is an intervention used in conservation agriculture system intensification (CASI) to enhance soil organic carbon (SOC) storage and other soil quality parameters as compared with conventional tillage (CT) system. No tillage operation eliminates all pre-planting mechanical seedbed preparation thereby reducing burning of fossil fuel and reduces sequester of carbon in the soil that ultimately reduces GHGs emissions. Intensive tillage operations involved in conventional agriculture creates the problems of soil degradation and oxidation of soil organic matter (SOM). Powlson *et al.* in 2016 reported that the rate of SOC increase under ZT in the tropical Indo-Gangetic Plains (IGP) could be around 0.3 Mg C ha^{-1} yr^{-1} in a 0-30 cm soil layer. As per the estimation by Grace *et al.* (2003), the CO_2 emission from CT field of rice-wheat cropping system could be about 29 Mg CO_2 per year if applied to 1 million hectares of IGP; while this amount can be lowered to 14 Mg CO_2 per year under ZT system with residue retention on the soil. As per the report by Grace *et al.*, (2012) in India, rice-wheat system has the potential to sequester 44100 Gg C over 20 years under ZT practice. Aryal *et al.* (2015) revealed that adoption of ZT under wheat cultivation reduced the GHG emission by 1.5 Mg CO_2 eq ha^{-1} yr^{-1}. Direct seeded rice (DSR) is the best alternative for conventional puddle transplanted rice that emits huge amounts of methane (CH_4) and other GHGs. Pathak *et al.* (2013) observed that direct seeded rice has the global warming potential of 1.3 - 2.9 t CO_2 ha^{-1} while it is 2.0 - 4.6 t CO_2 ha^{-1} for transplanted rice. Burning of crop residues can be reduced efficiently by CA practices. That means, it also lowers the emission of CH_4 and N_2O (each ton of crop residue burning emits 2.3 kg of CH_4 and 40 g of N_2O) (Grace *et al.*, 2003).

Direct seeded rice has potential to minimize the cost of production, soil health hazard, GHG mitigation and the negative impacts on the succeeding crops. Aerobic condition or intermittent wetting and drying reduce CH_4 emission from rice fields. Crop residue retention on the soil under zero-tillage is more beneficial than residue incorporation into the soil under conventional tillage. It regulates soil temperature, reduces weed infestation, conserves soil moisture and organic carbon, supplies nutrients, and reduces irrigation need, increases crop yields and net return. Crop residues help to enhance water infiltration, reduce evaporation loss and water erosion. Residue management will avoid hazards generated due to straw burning, improve soil organic carbon, enhance input-use efficiency and have the potential to mitigate GHGs emission. New variants of zero-till seed-cum-fertilizer drill such as turbo seeder, happy seeder and rotary-disc drill are used for direct drilling of seeds in the presence of surface residues. In India, about 500 Mt of crop residues are produced every year both on-farm and off-farm. The amount of surplus crop residues in India is 90-140 Mt which is burnt on-farm every year (Pathak *et al.*, 2011). With the proper application of CA technologies, the crop residues can be used for reducing pollution, improving soil health, increasing crop productivity and enhancing sustainability and resilience of agriculture.

8.4 Quantification of mitigation potential of conservation agriculture practices

Conservation agriculture practices mainly aim on two modifications: (i) minimal soil disturbance and (ii) retention of plant residues on soil. These two factors make positive changes in soil properties and physico-chemically as well as biologically processes. Under absence of intense soil disturbances due to tillage operations, moisture retention improves, soil aggregates are stabilized, organic matter is protected and microbial communities are less disturbed. Increased resistance to gaseous diffusivity under zero tillage causes low mobility of gases along the soil profile affecting gaseous transport. Under reducing conditions, incorporation of crop residue makes soil more anaerobic as organic compounds serve as electron donors. However, CH_4 emission rates are generally higher in no-tilled condition than in continuously tilled soil and N_2O and CO_2 emission rates are higher in alternately tilled soil due to crop residue incorporation coupled with intermediate soil disturbance. Alteration of drainage regimes and crop residue incorporation in rice production system can reduce CH_4 emission. Crop management practices that increase SOC in semi-arid region generally reduce CH_4 emissions. However, wetting and drying of soil also enhances N_2O emission and soil carbon mineralization resulting reduction in net GHG mitigation potential. Zhao *et al.* (2016) in a study indicated that the higher N_2O emission could be mitigated by adopting no tillage (NT) in alkaline soil for long-term duration and less nitrogen fertilization application when compared to conventional tillage (CT).

Pathak *et al.* (2012) conducted a study in two regions broadly classified as the upper Indo-Gangetic Plains and the lower Indo-Gangetic Plains to assess the potential and cost of low carbon technologies. Punjab, Haryana and western Uttar Pradesh comes under the upper IGP, while eastern Uttar Pradesh, Bihar and West Bengal comes under the lower IGP. In both the regions GHG emission was calculated for rice and wheat crop using the InfoRCT simulation model. Twenty different technologies were selected in rice and ten in wheat to study their GHG mitigation potential in respective crops. In rice crop grown in the upper-IGP, maximum reduction in GWP has been observed in carbon sequestration technology where rice straw was used as a construction material. Zero tillage also has shown a high global warming potential (GWP) with reduction strength of 83.9%. In case of wheat, GWP reduction strength ranged from 6 to 204% in the upper IGP and 5 to 252% in the lower IGP.

A preliminary estimate shows that global warming potential of about 25 Mt CO_2 eq. is mitigated annually with the adoption of various RCTs in India (Table 8.3). Every year, Indian agriculture (agricultural practices, rice cultivation and crop residue burning) emits about 172 Mt CO_2 eq. (Pathak, 2015). About 15% of the GHG emission from agriculture sector is mitigated by adopting various mitigation technologies. However, there are some uncertainties in assessing the impacts of RCTs on GHG mitigation and climate change adaptation under different management and agro-climatic conditions. Incomplete and incorrect implementation of RCTs may even enhance GHG emission and create adverse impacts on adaptation.

Table 8.3: Estimates of area under various conservation agriculture practices and their impacts on greenhouse gas mitigation and yield

Greenhouse gas	Mitigation technology	Area under mitigation technology (Mha)	Emission in conventional practice (kg ha^{-1})	Emission in mitigation technology (kg ha^{-1})	Total mitigation (Mt CO2 eq.)	Yield (% of conventional practice)
Methane	Intermittent wetting and drying	7.5	100	30	4.7	95
	Dry direct-seeded rice	4.0	100	40	3.4	90
Nitrous oxide	Leaf colour chart	40.0	1	0.15	1.9	110
	Neem-coated urea	120.0	1	0.15	5.6	105
Carbon dioxide	Zero tillage with residue	2.5	2000	600	1.5	105
	Integrated nutrient management	25.0	2000	300	7.5	110

Climate and soil characteristics determine where no-till management can store carbon in soils and mitigate GHGs emissions. Arid regions can get the most benefit from conservation agriculture by achieving enhanced C sequestration and increased crop yield (Hunt *et al.*, 2020). Carbonell-Bojollo *et al.* (2021) reported that the tilled system had a substantially greater environmental cost than no tilled system. The soil under the conventional tillage was found to emit 67% more CO_2 than the soil under the conventional tillage in four year conservation experiment growing different crops in rotation indicating the environmental conditions influenced the CO_2 emissions. Feng *et al.* (2018) in a global study observed that no tollage (NT) or reduced tillage (RT) significantly could reduce the global warming potential (GWP) of CH_4 and N_2O emissions by 6.6% as compared to conventional tillage (CT). Plaza-Bonilla *et al.* (2020) reported that NT could be the key for enhancement of several ecosystem services for the benefits in terms of GHG mitigation if combined with best management practices adapted to the specific condition in NT soils. Alam *et al.* (2019) reported that non-puddled transplanting with crop residue retention was an effective GHG mitigation option in wetland rice production.

8.5 Climate change adaptation with conservation agriculture

The CA practices give many benefits for climate change adaptation including adaptation to extreme weather events such as excess rainfall and heat stress (Table 8.1). In rainfed and dryland ecosystems, CA systems improve soil organic carbon and soil biodiversity; reduce soil erosion with crop residue cover and improve soil health by reducing tillage and adding legumes in the crop rotation. With improved soil organic carbon and increased water retention capacity crops can survive in mid-season drought in maize-horsegram system (Kundu *et al.*, 2013) and in pigeon pea based cropping system (Indoria *et al.*, 2017) in Alfisols of Southern India. Due to incorporation on crop residues, soil loss was minimized in rainfed pigeonpea-castor cropping system, besides improved soil moisture storage capacity in semi-arid tropics (Pratibha *et al.*, 2015, 2016). Other co-benefit of CA in rainfed cropping system is the reduction of nutrient and soil loss due to high intensity rainy events. Different studies on effect of long-term tillage operation and sources of nitrogen on crop yields of sorghum - sunflower crop rotation and soil carbon sequestration in rainfed vertisol also showed the similar advantage of climate change adaptation during variable rainfall situations in North Karnataka (Prasad *et al.*, 2013). Fruit-based cropping systems are the most suitable for sustainable production as they conserve soil and water and bring stability in crop production and farmer's income. Among the different intercropping systems tried, mango + guava + cowpea system improved the water holding capacity of soil to the extent of 30% and also improved organic carbon content in soil.

Climatic variability, especially rainfall variability (both drought and excess rainfall) and end season high temperature severely impact wheat production in India. Many studies have shown negative effects of high temperature during maturity stage of wheat in India (Gupta *et al.*, 2010). A large number of studies indicate that in irrigated production systems of India, CA based management practices help to enhance the resilience of agriculture to climate variability through better adaptation to climate change. A recent study in Haryana, India by Aryal *et al.*, 2016 showed that CA copes better with extreme weathers than the conventional tillage based wheat production system (CT). The study found that wheat yield during the bad year (defined in terms of untimely excess rainfall during grain development stage) was 4.89 Mg/ha under CA while it was only 4.23 Mg/ha under CT.

8.6 Conclusion

Conservation agriculture offers a new path for agricultural research and development to achieve food production target in India while solving the problems of natural resource degradation. Developing and promoting CA techniques on wider scale is highly demanding for the benefits of farmers and to ensure food security. However, the maximum benefits of climate change mitigation can be achieved by linking CA as an efficient and sustainable livelihood strategy to deal with degradation of the ecosystem. Standardization of equipment for sowing, fertilizer application and harvesting operation to ensure less soil disturbance under residue cover will be a key to the success of CA. There is also a need for policy analysis to obtain the knowledge on how CA technologies can be integrated with other technologies. Strengthening of knowledge and information sharing mechanisms are highly essential. Educating the farmers regarding the harmful impacts of residue burning is crucial. To achieve the sustainability in CA practices, it is essential to institutionalize CA into relevant government departments and regional institutions. For the successful adoption of CA at village level, provision of credit through banks and credit agencies at reasonable interest rates to farmers to purchase inputs, equipment and machinery is very much essential. More local support is needed to adapt and validate the CA technologies in rainfed cropping systems. Partnership of scientists, extension agents, farmers, policy makers and other stakeholders in the private sectors is important to develop and promote CA techniques at local, national and global levels.

References

Jat ML, Chakraborty D, Ladha JK, Rana DS, Gathala MK, McDonald A and Gerard B (2020). Conservation agriculture for sustainable intensification in South Asia. Nature Sustainability 3: 336–343

Pathak H, Sankhyan S, Dubey DS, Bhatia A and Jain N (2013). Dry direct-seeding of rice for mitigating greenhouse gas emission: Field experimentation and simulation. Paddy and Water Environment 11(1): 593–601.

Srinivasarao Ch, Gopinath KA, Prasad JVNS, Prasannakumar and Singh AK (2016). Climate Resilient Villages for Sustainable Food Security in Tropical India: Concept, Process, Technologies, institutions, and Impacts. Advances in Agronomy 140(3): 101-214.

Srinivasarao Ch, Prasad JVNS, Osman M et al. (2017). Farm Innovation in Climate Resilient Agriculture. Central Research Institute for Dryland Agriculture, Hyderabad, 123p.

Srinivasarao Ch, Ravindra Chary G, Mishra PK et al. (2014). Rainfed Farming – A Compendium of Doable Technologies. All India Coordinated Research Project for Dryland Agriculture, Central Research Institute for Dryland Agriculture, Indian Council of Agricultural Research, Hyderabad, 194.

Srinivasarao Ch, Venkateswarlu B, Lal R et al. (2011b). Long-term manuring and fertilizer effects on depletion of soil organic carbon stocks under pearl millet cluster bean–castor rotation in western India. Land Degradation and Development.

Srinivasarao Ch, Venkateswarlu B, Lal R et al. (2012a). Soil carbon sequestration and agronomic productivity of an Alfisol for a groundnut based system in a semi arid environment in South India. European Journal of Agronomy 43: 40-8.

Zhao X, Liu SL, Pu C, Zhang XQ, Xue JF, Zhang R, Wang YQ, Lal R, Zhang HL and Chen F (2016). Methane and nitrous oxide emissions under no-till farming in China: a metaanalysis. Glob Chang Biol. 2016 Apr; 22(4): 1372-84.

9

Rainwater Harvesting as an Alternative Water Supply in the Future

9.1 Introduction

Water is essential for all life and used in many different ways. It is an important part of the ecosystem in which the reproduction of the bio diversity also depends. Fresh water scarcity for agriculture, industry and household purpose is not only limited to the arid climatic regions, but also in areas having high rainfall. Water scarcity is caused by low water storage capacity, low infiltration rate, high evaporation demand and larger inter annual and annual fluctuations of precipitation.

The term water harvesting was first time used by Geddes of the University of Sydney. He defined water harvesting as the collection and storage of any form of water either runoff or creek flow for irrigation purpose. Meyer's of USDA, USA has defined it as the practice of collecting rainfall water from an area treated to increase runoff from rainfall. Recently Currier from USA has defined it as the process of collecting natural precipitation from prepared watershed for beneficial use. Now a days water harvesting has become a general term for collecting and storing runoff water or creek flow resulting from rainfall in soil profile both over surface and under surface. Previously water harvesting was used in arid and semi-arid regions, but recently their use has been extended to sub humid and humid regions too. In India water harvesting means utilization of the erratic monsoon rain water for raising good crops in dry tracks and conservation of the excess runoff water for drinking and for ground water recharging purpose.

When water harvesting technique are used for runoff farming, the storage reservoir will be soil itself but when the water is to be used for livestock, supplementary irrigation or human consumption, a storage facility of some kind will have to be produced. In countries where land is abundant, water harvesting involves: harvesting the entire rainwater, store it and utilize it for various purposes. In India, it is not possible to use the land area only to harvest rainwater and hence water harvesting means use the rainwater at the place where it falls to the maximum and the excess water is collected and reused in the same area. Therefore the meaning of water harvesting is different in different areas and countries. The water harvesting methods are used for both agriculture purpose and to increase the ground water availability.

The water harvesting for household purposes is also in existence for long years in the world. During rainy days, the people in the villages used to collect the rooftop water in the vessels and use this water for household purposes including drinking. In South East Asian countries, people used to collect the rooftop water by placing big earthen pots in the corners of their houses and they use this water for all household purposes. The main building of the Agricultural College at Coimbatore was constructed 100 years ago and they have collected all the rooftop water by pipes and stored in big underground storage tanks by the sides of the building.

Hence rainwater harvesting system is as old as civilization and has being practiced continuously in different ways for different purposes though out the world. The only concern is that it has not been done systematically in all the places. Now need has come to harvest the rainwater to solve the water problems not only in the arid but also in the sub-humid and humid regions.

Water harvesting was practiced more than 1000 years back in South India by construction of irrigation tank, temple tanks,farm ponds etc. In modern era the work is taken up at ICRISAT, Hyderabad, Central arid Zone Research Institute, Jodhpur, Central Research Institute for dryland Agriculture (CRIDA), Hyderabad, State Agricultural Universities and other dry land research centers throughout India.

9.2 Rainwater Harvesting System

Rainwater can be captured and stored by using rainwater harvesting system. Generally, rainwater harvesting refers to the direct collection of rainwater from roofs or other man made grounds; and the collection of sheet runoff from natural surface of catchment areas for domestic, industry, agriculture and environment use. Water harvesting systems can be classified as small, medium and large scale (Gould, 1999). Normally, the size of rainwater harvesting depends on the size of catchment area (Thamer *et al.*, 2007). The different categories of rainwater harvesting system are mentioned in Figurere 9.1.

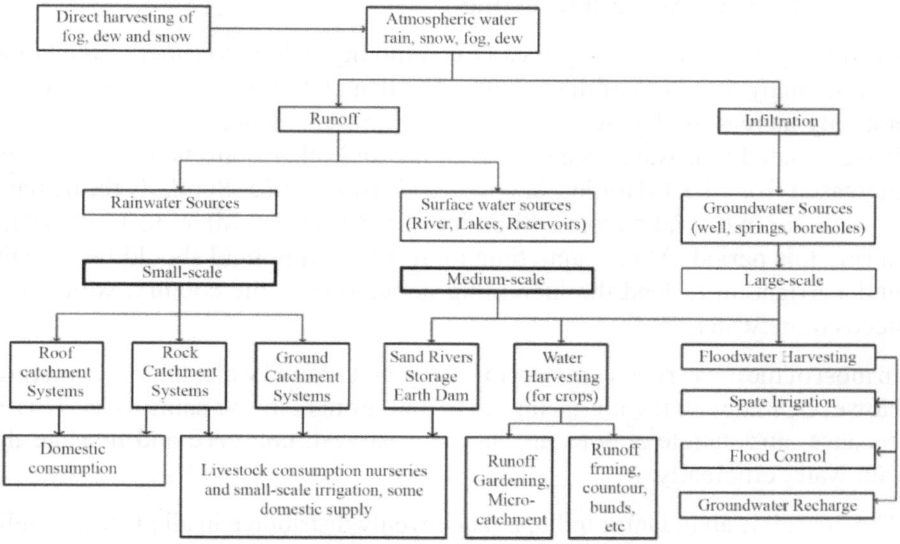

Figure 9.1: Rainwater harvesting category (Jamalluddin and Huang, 2007)

Rainwater harvesting system has been implemented and popularised in many developing countries like USA, Japan, China, India, Germany and Australia to support the increasing water demand for agriculture, industry and other sectors.

There are six main elements in rainwater harvesting system as mentioned in Figurere 9.2. They are catchment area, gutter and downspout, filtration system, storage system, delivery system and treatment/ purification.

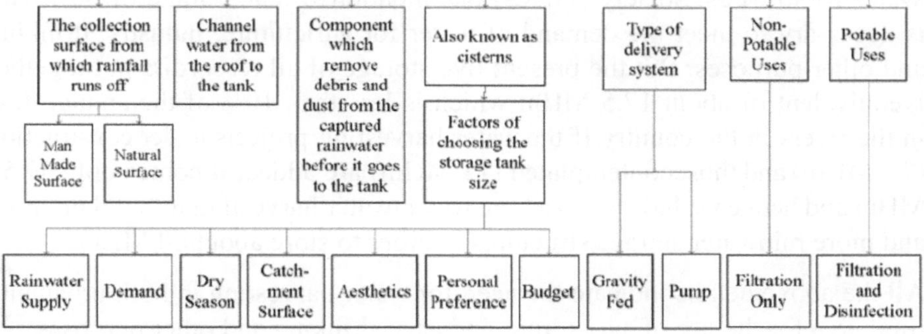

Figure 9.2: Elements of rainwater harvesting system

9.3 Need for rain water harvesting

Day by day water is becoming a scarce commodity andit is considered as a liquid gold in many dry areas of the country. The demand of water for Agriculture is not only increasing day by day, but also for household and industrial purposes. It is estimated that water need for drinking and other household uses will be increased from 3.3 MHm in 2020 to 7.00 MHm in 2025. Similarly the demand of water for industrial sectors will be increased from 3.0 MHm to 12.00 MHm during this period. At the same time more agriculture field should be brought under irrigation to feed the increasing population of the country, which also needs more water.

In most of the areas rivers are becoming dry and ground water table is depleting. However, many states are facing flood and drought in the same year. This is because, no concrete actions are taken to harvest, conserve and manage the rain water efficiently.

The rainfall is abundantin India, but not evenly distributed in all places.In India rainfall depends on monsoon and occurs only for 3 to 4 months in a year with high intensity, whichresults more runoff and soil erosion. It also erratic and fails once in 3 or 4 years which is very common phenomena in many parts of the country.

Water scarcity is not only limited to the arid regions, but also occurring in high rainfall areas also. Chirapunji of Meghalaya gets more than 11,000 mm of annual average rainfall but face drinking water problem before monsoon whereas in Ralegoan and Siddhi of Maharastra there is no water scarcity though the annual average rainfall is only 450 mm. Hence to mitigate water problem, stress, drought etc., there is an urgent need to follow our ancestral way of water harvesting and to adopt the latest technologies in soil and water conservation measures on watershed basis.

The Theme paper on Water vision 2050 of India, prepared by Indian Water Resources Society (IWRS) has mentioned that storage of 60 MHm is necessary to meet the demand of water for agriculture, industry, drinking and other purposes. But the present live storage of all reservoirs put together is equivalent of about 17.5 MHm which is less than 10% of the annual flow in the rivers in the country. If the water harvesting projects under construction (7.5 MHm) and thosecontemplated (13 MHm) are added, it comes only 37.50 MHm and hence we have to go a long way inwater harvesting to build up more and more rainwater storage structures in order to store about 60 MHm.

All the above details indicate the need for water harvesting in both urban and rural area for the use of agriculture, industry, drinking and other purposes.

9.4 Traditional water harvesting techniques

A. Ponds / tanks: These are the most commonly used methods to collect and store rain water. Most of the ponds have their own catchment areas which provide the requisite amount of water during the rainy season. Where the catchment area is too small to provide enough water, water from nearby stream is diverted through open channel to fill the pond. In some places water from irrigation canals is also used to fill ponds.

Ponds are excavated in different shapes and sizes depending upon the nature of the soil, availability of land, water requirement of the village community etc. These ponds are known by different names in different regions. In someareas, the ponds are called by names as follows:

- Nagaland – Zabo
- Gujarat – Kunda, Jheel
- Orissa – Katas, Mundas
- Maharashtra – Bandharas
- Karnataka –Volakere, Katte or Kunte
- Andhra Pradesh – Tank (Mainly for irrigation)
- Kerela – Tank
- Ladakh – Zing
- Jammu – Chhapris
- Sikkim – Khup

Figure9.3: Village pond for rain-water harvesting

B. Ground water harvesting: In hilly areas of Uttarakhand, the people harvest ground water by making stonewall across ground water stream. These are called Naula or Hauzi. For various reasons, there has been a steady decline in the construction of these structures, the main reason being drying up of underground streams due to large-scale deforestation and increased human activity in the hills. Similar practice is in vogue in parts of Kerela where the ground water is collected by excavating long deep trenches across a gentle slope. These are called Surangam. In Punjab, shallow wells dug near the streambeds to trap seepage water are called Jhalars. In Rajasthan these wells are called Beris which were were built by kings in earlier times, but now these are not in vogue. In Gujarat shallow wells called Virdas are dug in depressions to tap ground water. In Tamil Nadu Ooranis were built earlier but are not built any more.

C. Hill slope collection: In this system, which is in vogue in many hilly areas with good rainfall, lined channels are built across the hill slopes to intercept rain water. These channels supply water for irrigating terraced agricultural fields. The water is also stored in small ponds for domestic use and the cattle. These practices are seen in Himachal Pradesh, Uttaranchal, Meghalaya and Arunachal Pradesh.

Figure 9.4: Baori in Rajasthan to conserve rain water

9.5 Modern structures for water harvesting

A. Roof top harvesting: This system is useful mainly for drinking water purposes. In this system, rain water falling on roofs of houses and other buildings is collected through a system and is stored in tanks located on the ground or underground. The practice is in vogue at the individual household level in remote hilly areas with high rainfall and also in some semi-arid areas in the plains. This system can be seen in the northeastern states of Arunachal Pradesh, Assam, Meghalaya, Manipur and Nagaland. This is also in use in Bikaner, Jaisalmer and Jodhpur districts of Rajasthan. In recent years, at the initiative of the Central and State Governments, the practice has been increasingly adopted in many cities and towns in different parts of the country.

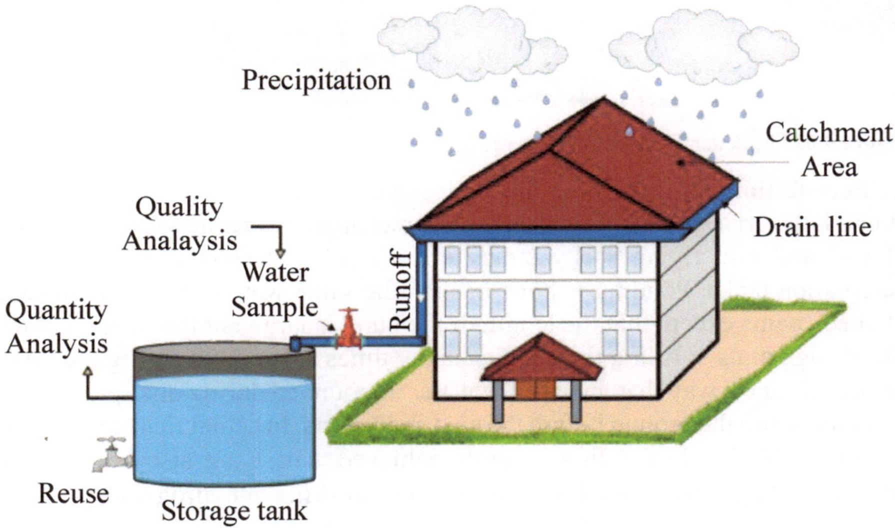

Figure 9.5: Schematic representation of roof top rainwater harvesting

B. Check Dams: These are concrete or masonry structures built across small streams for surface storage and incidental benefit of ground water recharge. The design of these structures are done taking into consideration the volume of water that can be stored in the stream channel upstream, the surplus flood discharge that must be evacuated safely, stability of the structure against various forces and the likely ground water recharge.

These are usually built by the State Government agencies like the departments of Irrigation/ Water Resources, Agriculture and Forests. These are the modified and improved versions of the traditional temporary or semi-permanent structures that people in the villages usually build across natural streams or drainage channels.

Figure 9.6: Check dam for rainwater harvesting

C. Percolation Tanks: These are built mainly to impound monsoon runoff over a large area to augment ground water recharge. Moderate to high porosity of soil and underlying rocky strata is the main criteria for the choice of percolation tanks. Ponding is done in such the same way as it is done in case of check dams except that the length of the bund is large but the height is low. The design aims at filling the pond as many times as possible during the rainy season in such a way that most of the water impounded during one spell of rain percolates into the ground before the next spell starts. In actual field conditions, however, this ideal operation is rarely achieved. These are also built by the government agencies since these require special skills in hydrogeology.

Figure 9.7: Percolation tank in Yavatmal of Maharashtra
Source: ISC report

D. Sub-Surface Dykes: These are impermeable walls or barriers in masonry, concrete and/ or clay built below the bed level across natural streams to arrest sub-surface flow of water to improve the yield of existing wells and hand pumps in the upstream.

Besides these direct methods of water harvesting some indirect methods have also been developed. These aim at augmenting soil moisture retention and preventing soil erosion and land degradation. These are:

(i) **Contour Bunding:** These are small earthen bunds built horizontally in parallel rows across the hill slope. These help in maintaining soil moisture and prevent erosion of topsoil.

Figure 9.8: Rice cultivation in contour bunds

(ii) **Gully Plugging:** These are soil and water retaining structures built across gullies in hilly areas. These are built with locally available materials like stone boulders, earth, brushwood etc.

Both contour bunding and gully plugging are the part of watershed improvement works. The other works in this category are:

- Bench Terracing
- Contour Cropping
- Contour Trenching

Figure 9.9: Gully plugging for rainwater harvesting

9.6 Case studies of rain water harvesting

Though there are many sporadic studies, but some studies that need special mention are given below for purpose of reference and information:

- Watershed management programme launched by the Government of Madhya Pradesh in 1995 rising water levels in many areas of the state. More than 1000 check dams, 1050 tanks and 1100 community lift irrigation schemes have been implemented in Jhabua district of MP. Food production of this district has increased by 38% in the past five years.
- A micro-watershed project implemented in Ghelhar Choti village of Madhya Pradesh has led to recharging of ground water. Because of this, the cultivate area has increased and the yield per ha has doubled.
- District Rural Development Administration in Rajkot implemented 50 micro-watershed projects under the watershed management project launched by the Govt. of Gujarat in 1995-96. Rise in ground water levels has been reported from wells in areas, where these projects have been implemented.
- More than 70,000 percolation tanks have been built in Maharashtra after the severe drought in 1971-72. All such small catchments of percolation tanks have been converted into green patches.
- Underground bandharas have been constructed in various parts of Maharashtra, viz. 87 tehsils of DPAP areas with 400-700 mm annual rainfall located in parts of Sangli, Satara, Pune, Sholapur, Ahmednagar, Nashik, Aurangabad, Usmanabad, Beed and Buldana districts; Vidarbha

and Amravati divisions where there are no surface sources and parts of Konkan area with high well density. Assessment of benefits from recharged ground water indicates that cost benefit ratio would not be less than 1:1:5.

- Sadguru Water & Development Foundation has constructed a number of concrete check dams in Dahod district of Gujarat with the help of local residents to impound rain water for irrigation in Thunthi Kankasiya village. The Foundation has also implemented watershed management measures such as trenching, bunding, terracing and planting trees. The average household income has been raised from Rs. 8,000 per year to Rs. 35,000 per year.
- Saurashtra Jaldhara Trust is working at 100 villages in Amravati, Bhavnager, Jamnagar, Junagarh and Rajkot districts. In Khopala village of Bhavnagar district the trust has motivated the villagers to build rain water harvesting structures and the villagers have built 210 low cost check dams on streams in and around the village.
- Raj-Samadhiyala village in Rajkot district, which was once declared a desert area is not a water scarce village now. This has been possible due to watershed management project taken up by the villagers. The villagers built 12 check dams between 1986 and 1988. This has brought prosperity and social wellbeing to the village.
- Villagers of Gandhigram village in Kutch district of Gujarat have been facing water crisis for the past 10 years. The villagers have constructed a dam on Khari river and have undertaken a micro-watershed project. With the help of rain water harvesting villagers are able to cultivate crops even during drought and the crop production has also increased several time.
- The Dewas district administration of Madhya Pradesh has banned tubewell drilling and made roof top rain water harvesting mandatory for all houses. Fresh ponds were also constructed to maintain water level during summers. This has helped in improving the moisture content in soil and recharging shallow aquifer.
- Chennai Metro Water Board has made it mandatory under the city's building regulations for all new buildings to have water harvesting mechanisms to recharge ground water aquifers.
- The administration of Delhi has made roof top rainwater harvesting mandatory for at least in new buildings measuring 500 sq.ft. or more.

9.7 Importance of Rainwater Harvesting

Rainwater harvesting is an important part of sustainable architecture and it is bring a lot of advantages not only for the users but also to environment and government. In short, Table 9.1 shows some of the benefits from rainwater harvesting system.

Table 9.1: Benefits of rainwater harvesting to users, environment and government

Users	Government	Environment
• Rainwater is a source of relatively clean, independent and ample water supply. • For rainwater harvesting simple technologies are used that are inexpensive and easy to maintain. • Rainwater harvesting system is very easy to handle. It can be modular in nature by allowing expansion, reconfiguration or relocation. • Rainwater harvestingsaves money by reducing the volume of water purchased from outside or public systems. • Save money by extending the working life of plumbing fixtures and appliances. • Avoid interrupted service from centralized water systems or over use of water from wells.	• Reduce the burden for new investment to replace the ageing systems and adding the water supply infrastructures. • Potentially avoid the cost of accessing public water systems when it is not economically feasible. • Rainwater harvesting reduces construction cost in each development because it can be easily retrofitted to anexisting structure or built a new construction.	• By capturing rainwater, wereduce the abundant amount of rainwater that goes to the drainage and avoiding the floods phenomena. • We can significantly reduce our reliance on water storage dam. This will avoidsecological damage to the area which has to be submerged to build the dam. • Increase soil moisture level for agriculture. • Rainwater is superior for landscape use and plants thrive on rainwater well rather than other sources thatmight have chemical compounds that polluted them and soils. This is because rainwater forces salts down and away from root zone when it percolates into the soil

9.8 Conclusion

It is very important to make water everybody's business. It means a role for everybody with respectto water. Every household and community soul involve in the provision of water and in the protection of water resources. Make water as the subject of a people's movement. It means the empowerment of our rural and urban community and to manage their own affairs with the state playing a critical supportive role.

Further involvement of people will give the people greater ownership over the water project including watershed development. Soil and water conservation and water harvesting will go a long way towards reducing misuse of government funds. It will also develop the ownership of own water supply systems. Inthis way it is possible to solve water problems facing by the county in the 21st century.

References

Ake Nilsson. 1988. Ground water dams for small-scale watersupply, IT publication.

Center for science and environment. 2001. A water-harvesting manual, Delhi.

Center for Science and Environment. 2001. Making water everybody's business, New Delhi.

Chitale M A. 2000. A blue revolution, Bhavans BookUniversity, Pune.

CII. 2000. Rainwter harvesting – A guide, New Delhi.

Gould J. 1999. Contributions Relating to Rainwater Harvesting. Paper prepared for the World Commission on Dams Secretariat (WCD) Thematic Review IV.3.

Jamaluddin S A and Huang Y F. 2007. NAHRIM's Experince in Rainwater Utilisation System Research. Paper presented at Rainwater Utilization Colloquium on 19 & 20 April 2007 at NAHRIM Mini Auditorium.

Rajiv Gandhi. 1998. National Drinking water missions Handbook on Rainwater harvesting, Government of India, New Delhi.

Sivanappan R K. 1999. Soil and Water Conservation and Water harvesting, Tamil Nadu Afforestation project, Chennai.

Sivanappan R K. 2001. Water harvesting, ICCI, Coimbatore.

Stockholm Water Symposium. 1998. 'Water harvesting' Stockholm.

Thamer A M, Megat-Johari M M and Noor A H G. 2007. Study on Potential Uses of Rainwater havesting in Urban Areas. Paper presented at Rainwater Utilization Colloquium on 19 & 20 April 2007 at NAHRIM Mini Auditorium.

Verma H N and Tiwan K N. 1995. Current status and Prospects ofRain Water Harvesting, NIH, Roorkee.

